The Hidden
Geometry of Life

The Hidden Geometry of Life

The science and spirituality of nature

Karen L. French

To my family

The Hidden Geometry of Life
Karen L. French

This edition first published in the UK and USA in 2015 by
Watkins, an imprint of Watkins Media Limited
Unit 11, Shepperton House
89 Shepperton Road
London N1 3DF

enquiries@watkinspublishing.com

Managing Editor: Sandra Rigby
Produced by Bookworx
Editor: Jo Godfrey Wood
Designer: Peggy Sadler

A CIP record for this book is available from the British Library

ISBN: 978-1-78028-108-7

10 9 8 7 6 5 4

Colour reproduction by Imagewrite
Printed in Malaysia

www.watkinspublishing.com

Contents

Introduction 6

PART I Shaping Reality 8
1 Language of Numbers and Their
  Shapes 10
2 Circle, Unity and the Duals 18
3 Square, Space-Time and the Cross 28
4 Triangles and Being 34
5 Matrix of Space-Time-Being 42
6 Spiral Life Force 46
7 Numbers 5 and 10 54
8 Pattern-Sharing and Fractal
  Geometry 62

PART II Structure of Being 72
9 To Be and Being 74
10 Mind and Sentience 79
11 Classical Elements as Metaphors 88
12 Classical Elements and the Platonic
  Solids 96
13 What is Matter? 102
14 Geometry in Multiple Dimensions 112

PART III Medium of Sound 130
15 Sound as a Vehicle for Geometry 132
16 Sacred Sounds 142
17 Music, Rhythm and Harmony 148

PART IV Let There Be Light 158
18 Light Beings 160
19 Colours of Light 170
20 Music and Colour 180

PART V Gateway to Becoming 188
21 Holographic Universe 190
22 Illumination and Intention of
  the Mind 193
23 Gateway to the Heavens 197
24 Earth 201
25 Water 204
26 Fire 208
27 Air and Ether 213
28 Living Art 216
29 Combining Shape, Colour and
  Element 220
30 Gateway to Becoming 224

Glossary 232
Index 236
Picture credits 239
Acknowledgments 240

Introduction

MANY YEARS AGO A DOOR OPENED AND I WAS DRAWN into the fascinating language of numbers and their shapes. It became part of a very special journey for me; a journey that has led me down numerous roads of research into a wide variety of subjects, taken me to different countries and introduced me to many fascinating people. Out of these experiences *The Gateway Series* began to take shape. First was the *Gateway to the Heavens* and now, continuing with my journey, I would like to introduce you to my second book, *The Hidden Geometry of Life*.

This book is the culmination of years of research and the experience of using geometry in many different aspects of my life; be it in business and the organization of information, my art, in collating research and generally in how I view the world around me. I include personal insights and interpretations, backed up by the insights of current and earlier generations across different cultures and I have been able to introduce some amazing scientific models that have only recently been put forward. Pertinent information from *Gateway to the Heavens* is included, added to and organized differently, as it is now placed in a different context.

My intention is to provide an overall insight into this fascinating subject by distilling information from a wide range of areas and showing how various pieces of the puzzle fit together. Because of this I can only show you the surface aspects and key points of some very complex topics, such as fractals and yantras, but I hope that *The Hidden Geometry of Life* stimulates you to delve into greater detail for yourself.

My first book, *Gateway to the Heavens*, is about the meaning and purpose of key constituent parts of reality, made comprehensible through a few basic geometric shapes and principles. These are, essentially, the Square, two Triangles, a Circle, Cross and Spiral. As I explain in that book, those shapes are expressed in the symbol known as the Sri Yantra (see the illustration on the facing page), which translates as "gateway to the heavens". These shapes have no colour, no substance; they are perfection in form, providing grids to form a Matrix on which space, time and Being are moulded. The Square represents space, the Triangle life and the Circle time. These same shapes can then be exploited as powerful symbols and tools, which you can use to expand awareness of your reality, since they structure it.

But reality is not sterile and monotone, comprising spheres, boxes and lines. From a few basic geometric shapes and rules the endless variety, sophistication and diversity of the myriad forms of Being come into existence. Nature's palette has flavour, diversity, vibrant colour and rhythm to stir body and soul. Amazingly, although humans are minuscule entities in terms of the vast scale of the cosmos, we have evolved as Beings with the valuable capacity to reflect and question our situation; what we see, feel and experience beyond mere survival.

I wanted to understand more about what it is "to be" and the process of creation, such that the infinite variety and hue of forms of Being can "become" and individualize, yet still co-exist and be part of the cosmic, interconnected unity; I wanted to know how the geometric blueprint is accessed, used and transferred in the process of creation. I discovered that some of the answers revolved around the roles and purpose of sound, light and the Classical Elements of earth, fire, air and water. Central to these considerations is how these are combined with the formal, structural Gateway to the Heavens model, so that it is transformed into a vibrant, creative Gateway to Becoming model, assisting the production of boundless variety.

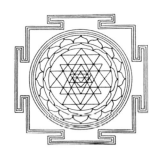

Above **The Sri Yantra mandala.**

Underpinning everything is the amazing evidence provided by both the natural and man-made worlds that reveals how patterns, shapes, sound and light provide us with insight into the biggest questions we ask ourselves: What is reality? Why am I here? Is there meaning to life? I want to show you, through my own understanding of geometry and spiritual insights, how the answer is very much in the affirmative – there is meaning and we can all access the tools to better understand it. And finally I hope that you will be surprised at the results of combining geometry with science, art and spirituality.

How to use this book

This book has been conceived as a journey through geometry, sciences, the arts and spirituality. As you progress through each part, I will be building up a picture that combines these disciplines.

The book begins, in Part I, by looking at basic shapes such as the Square, the Triangle and the Circle from various perspectives, such as their structural purposes and symbolic significance, individually and together as a group. Moving on, in Part II more detailed ideas about the building blocks of matter as we understand it from an atomic and sub-atomic level are explored, and how Nature uses pattern-sharing to spawn a huge variety of life forms. Part III then goes on to explain how sound contributes to propagating reality as we experience it and Part IV looks at the vital role played by light. Part V brings all these concepts together and also explores unified ideas about space, time and life in the Universe, to show how insights from ancient and mystical traditions correspond to the latest scientific theories about the world in which we live.

A note on style

In this book you will notice unusual stylistic features. Particular words have been capitalized throughout, for example the name of key geometric shapes and patterns, such as the Circle, Square and Spiral, and important concepts such as the Duals. This has been done to give these names and terms special emphasis that reflects the degree to which they underpin my arguments through the book. In addition, sometimes I have chosen to express numbers as figures in order to draw your attention to the number itself, its own unique shape and the symbolic as well as practical power it holds.

> Creativity is not the finding of a thing, but the making something out of it after it is found.
>
> JAMES RUSSELL LOWELL, AMERICAN DIPLOMAT, ESSAYIST AND POET (1819–1891)

PART I

SHAPING REALITY

…and in the shifting of the winds, and in the clouds that are pressed into service betwixt heaven and earth, are signs to people who can understand.

KORAN

Your journey into the meaning and purpose of simple shapes begins here in Part I – Shaping Reality. You will learn how the Circle represents such concepts as the Void, the dynamics of relationships, unity and the passage of time, and a concept I call the Duals. You will discover how the Square represents our grounding in the physical world and the symbolic role of the Cross in emphasizing how the choices we make in life influence the directions in which we may go. You will see how the Triangle can structure different tiers of life that we may or may not be able to sense. Then you will be shown how a Matrix, built by a combination of the Circle, Square, Triangle and Cross, provides a model by which you can understand your place in time and space.

The Spiral is revealed as a symbol of the Life Force, which animates our world, and you will see how it relates to additional essential numbers and shapes, such as the pentagram. You will learn how patterns are shared, especially in fractal geometry, providing further order to add to your reality on every scale.

1 Language of numbers and their shapes

Above *Melancholia* by Albrecht Dürer (1471–1528) shows Melancholy. Her wings represent the flight of her imagination and yet she sits idly since she is lost in thought. Keys and a money bag symbolize power and wealth. She is surrounded by measuring instruments and geometric solids, while at her feet are the tools that can style the material world.

There exists, if I am not mistaken, an entire world which is the totality of mathematical truths, to which we have access only with our mind, just as a world of physical reality exists, the one like the other independent of ourselves, both of divine creation.

CHARLES HERMITE, FRENCH MATHEMATICIAN
(1822–1901)

MANY THOUSANDS OF YEARS AGO HUMANS DEVELOPED AN INCREDIBLE GIFT: the ability to imagine. Being able to imagine gave us the faculty with which to question our surroundings and to form images in our minds. We started to record these images as sketches on the walls of caves, on pieces of rock and other surfaces. From these humble origins we have evolved many complex symbolic systems in the arts, sciences and faiths, but there remains an underlying consistency in the meaning and use of numbers and geometric shapes across wide-ranging cultures and eras.

It is this development of the use of numbers and shapes that is so extraordinary. Our insights were not just based on the obvious surface appearance of our surroundings. For example, the Circle and the Spiral are both obviously prevalent everywhere in the natural world, but what about the Square and numerous other shapes and patterns, such as the pentagon, the hexagram and the dodecahedron? How often do you see a perfect cube in the natural environment? It is amazing that in our imagination we can intuitively appreciate that Squares and cubes provide straightforward order and act as simple, rigid containers that hold things; which is why we use them so prolifically in designs of buildings, road layouts, boxes and even our vocabulary to describe the experience of being "boxed in".

"We must admit with humility that, while number is purely a product of our minds, space has a reality outside our minds, so that we cannot completely prescribe its properties a priori."

CARL FRIEDRICH GAUSS,
GERMAN MATHEMATICIAN AND PHYSICIST (1777–1855)

Left In this 1888 woodcut by Camille Flammarion (coloured by Hugo Heikenwaelder, Austria, 1988), a man looks through the Earth's atmosphere, as if it were a veil, to examine the inner workings of the Universe.

Above From a young age our imaginations are starting points for exploration of the world. Simple geometric shapes are among the first concepts children learn to draw.

The first marks of humankind

Carvings and scratchings of simple and complex shapes are found the world over, testimony to humankind's early record of its presence.

Oukaimeden, Morocco

Cundinamarca, Colombia

Grand Canyon, Arizona, USA

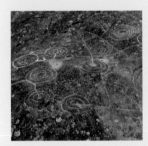

Marin, France

Caicara, Venezuela

Phoenix, Arizona, USA

Mali, Africa

Cholpon-Ata, Kyrgyzstan

Above Religious and metaphysical texts tell us that God created the Universe according to geometric and harmonic principles. To seek these principles was therefore a part of seeking and worshipping God. This drawing shows a Christian representation of God as architect, builder, geomancer and craftsman.

Above **Shapes and patterns are fundamental to Nature's designs. From atoms and cells to their final form, geometry makes flight and growth possible, enables water to flow and can be seen in the orbits of the planets.**

Inner knowing

Through words we invoke the images formed within us and share them with others; images that in turn stir the soul and emotions. Numbers have their own language, their own imagery and a life that can resonate strongly within us. Their language is translated into visual terms as the images of shapes and principles, such as duality, which become symbols, like the Spiral and the T'ai Chi. Numbers and their shapes may be products of our Minds, but their language, this imagery, is a living, permanent language celebrated throughout Nature, as seen in her creations. It remains vibrant and unaffected by the changing values of the prevailing society, so retaining its capacity to "wake up" each new generation, if it chooses to read it. This language is part of an archetypal inner knowledge that is both logical and intuitive and can be imagined.

Each shape and geometric pattern is like a word in a language. Each has its own meaning, but true significance and power emerge in recurring relationships and as dynamic patterns woven together in an ordered whole. Once the timeless language of shape and pattern is understood, reading and living with it are personally empowering in much the same way as enjoying great writing. It moves us in body and soul when we intuitively recognize how this language unites us with the Universe, since every part of us and our life is being moulded and driven by the same principles.

> Beauty or ugliness, order or confusion are only relevant in relation to our imagination.
>
> BARUCH SPINOZA,
> DUTCH PHILOSOPHER
> (1632–1677)

> Mathematics is the most beautiful and most powerful creation of the human spirit.
>
> STEFAN BANACH, POLISH MATHEMATICIAN (1892–1945)

Symbolism and deeper truths

Symbols are fundamental to our consciousness, thoughts, the nature of being human. They serve as vital codes we rely on to share experiences with others, so that they can see what we see and feel how we feel. Inextricably, symbols link our inner and outer worlds, making associations between thoughts and "reality" as we perceive it.

Beyond the limitations of their physical representation, symbols relate to deeper truths. To understand these truths requires the use of our intuition, since symbols provide us with insights that underlie the source of our creativity. The same creativity has facilitated our evolution because it has allowed us to see beyond the boundaries of the physical, the logical and the rational. True symbols "commune" with our hearts and souls; in effect they have a life of their own.

Above **From the Yin-Yang and the dove of peace to the character signifying AUM, the world is full of symbols. Simple and direct, they communicate their meaning in ways that words cannot.**

Below **A symbol for Chak, the Mayan God of rain and lightning.**

Sacred geometry and mundane geometry

Literally meaning "the measuring of the Earth", geometry is concerned with the mathematical rules and codes making up the blueprint of the Universe. Geometry has grown from the original insights of humanity about the order governing reality – the "limits that give form to the limitless". Shapes and patterns we call the Square, Circle, Triangle and Spiral are part of geometry. As archetypes we recognize them intuitively and use them as symbols and tools in direct relation to their structural purposes. Because of this, geometry is an intrinsic part of every facet of our lives and is at the core of the rich diversity of symbolic systems evident around the world. Accept the word "geometry" into your life and do not worry about the mathematics.

When we employ geometry in routine, mechanical ways such as for measurement and calculation in art and design, it is known as "mundane geometry". Many people deem all geometry to be sacred in recognition of its role in expressing what some religions and scholars call the Divine Plan and the blueprint of reality. Humankind's use of geometry to fashion our own creations, such as in art and architecture, is an integral part of our experience of reality. Being creative contributes to our learning and evolution, and to the evolution of the Universe as a whole. Intent, the motivation behind a thought, a word or a deed, is the key to activating geometric shapes and patterns. True sacred geometry utilizes intent to release the inherent power of the number or the shape in our creations.

Sacred geometry is recognized as a common global heritage employed by the Inca, Egyptians, Romans, Celts, Asian Indians, Japanese, Australian Aborigines,

Native Americans and the tribes of Africa. As sacred geometry grew in importance in many societies, geomancers, the experts in studying and employing sacred geometry in art and architecture, would have had a rounded education in mathematics, sciences, arts and the music of their time. It will become apparent why this was the case as this book unfolds and current advances in science, sound, colour and the dynamics of art are merged with geometry and the unified model called the Gateway to the Heavens, which is the key geometric model underpinning this book. Although each discipline has depth and gravity in its own right, when their contributions are intertwined they enrich each other and our appreciation of the whole.

> *"Philosophy is written in this grand book – I mean the Universe – which stands continually open to our gaze, but it cannot be understood unless one first learns to comprehend the language and interpret the characters in which it is written. It is written in the language of mathematics, and its characters are triangles, circles and other geometric figures, without which it is humanly impossible to understand a word of it; without these, one is wandering about in a dark labyrinth."*
>
> GALILEO GALILEI, ITALIAN PHYSICIST, MATHEMATICIAN, ASTRONOMER AND PHILOSOPHER (1564–1642)

> *"The mathematics are usually considered as being the very antipodes of Poesy. Yet Mathesis and Poesy are of the closest kindred, for they are both works of the imagination."*
>
> THOMAS HILL, UNITARIAN MINISTER, MATHEMATICIAN AND SCIENTIST (1818–1891)

Above **The title page of *Geometriae practicae novae et auctae tractatus*, a manual by Daniel Schwenter (1585–1636) for practical geometry and surveying, showing Pythagoras, Archimedes, a 3-4-5 Triangle and drafting and surveying instruments.**

Above **Illustration depicting the study of astronomy and geometry, showing an armillary sphere, a Square and compasses being held and on the table (1619).**

Above **A collection of traditional instruments used for drawing geometric forms.**

Sacred universal models in art

Beautifully constructed combinations of various shapes and patterns in geometric models are employed in cultures throughout the world in sacred rites, in art and in structures constructed as scaled-down representations of the celestial on Earth. The mandalas and yantras of Asia are precise works of art used in conjunction with toned sound to evoke the power of the image. Like the Medicine Wheel of the Native Americans and the Jewish Kabbalah, they have been conceived and created down the centuries to give tangible form to the intangible order that unites everything in the Universe.

Every culturally coloured model facilitates the sharing of elusive concepts and relates to the deeper truths of our experience of reality. Revealing the true nature of the Universe through abstract forms, their complex symbolism may be reduced right down to outwardly simple patterns and images that actually contain a great deal of information. The fundamental power that universal models based on geometry carry is an immediate, intuitive knowledge of the Universe and the "meaning of life". They are not just representations of the complete picture of reality; they are indistinguishable from reality.

Temple architecture
Evolving from standing stones based on the Circle, the use of sacred geometry in architecture has become integral to every aspect of the art.

Sainte-Geneviève, France

Mnajdra, Malta

Stonehenge, Britain

Al-Azhar, Egypt

St Basil's, Russia

Summer Palace, China

Prasat Hin Muang Tum, Thailand

Big Horn Medicine Wheel, USA

The essence of reality

Every geometric shape has a function and it holds together, or sustains, energy. Each shape provides boundaries for the unbounded and frameworks for the hugely complex, so that it can act as an aid to our perception of reality and its structure. The perception of a boundary, or edge, is created when we draw a shape using lines. In symbolic systems (and this includes the modern sciences), shape and what it signifies beyond a literal form are very important. A shape fashions an aspect of reality, but it is not real in a material sense, only symbolically.

Individual shapes can be studied separately, but complete comprehension requires an immediate appreciation of their fused whole. You may consider geometric forms just as something that is structural and technical, or perhaps you are averse to science and prefer to look at them as emotive and inspirational imagery. Each of us has our bias and favoured way of viewing the world. In this book complex ideas are simplified and condensed and it is important to remember that full appreciation of each shape and universal geometric models requires a merging of logic, creativity and intuition. Keep in your mind that shapes and patterns represent ideas and distil the essence of reality.

> Limit gives form to the limitless.
>
> PYTHAGORAS,
> GREEK PHILOSOPHER
> AND MATHEMATICIAN
> (c580–c500 BC)

Geometry in sacred artwork

Even on a flat surface sacred geometry has been, and still is, universally employed in both wonderfully simple and ornate designs.

Ancient Jewish geometric mosaic

Islamic art

Tibetan sand mandala

Christian cross

Sacred Buddhist lotus pedestal, Japan

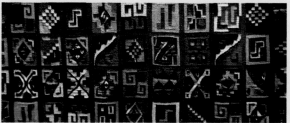

Tukapu patterns on Inca fabrics

Native American art

2 Circle, unity and the Duals

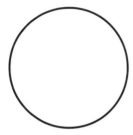

Above **The eternal Circle, holding all possibilities.**

Indefinable yet ever present, it is nothing at all.
It is the formless form, the imageless image.
It can't be grasped by the imagination.
It has no beginning and no end.
This is the essence of Tao.
Stay in harmony with this ancient presence,
and you will know the fullness of each
present moment.

LAO-TZU, CHINESE PHILOSOPHER (6TH CENTURY BC)

EVERYTHING BEGINS WITH THE CIRCLE, so it is the first simple shape we will examine. The Circle calls attention to the message it encircles. For example, when a teacher puts rings on schoolwork or when traffic signs emphasize an instruction. As the source of all subsequent numbers, and hence shapes, ancient mathematical philosophers regarded the Circle as the first number. Of infinite proportions, it can expand forever, providing a vast vessel in which the Universe can unfold. The Circle appears to be an empty void and is used as a "zero", to denote nothing, but actually it is a container of latent potential "to be", making it full and empty. Because of this it is the Absolute; "complete, perfect, pure and unrestricted".

Left to right **Prehistoric Chinese petroglyph; Capitol building, Washington, DC, USA; circular gateway framing an idyllic Chinese garden, Vancouver, Canada. Circles draw attention to the message and Centre.**

Circles great and small

Circles and spheres are the primary form of Nature and are seen everywhere, from minute delicate ash particles to vast planets.

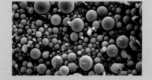

Microscopic ash particles

Above **Road signs employ the Circle to call attention to their message.**

Soap bubbles

Planet Earth

Craters on the surface of the Moon

Above **Organizing our lives is made easier when we use Circles to emphasize things we need to remember – for example in personal diaries and on calendars.**

Grapes

Grains of sand

Submarine volcano vent site

Monad

"Monad", in Greek, describes the Circle and originates from *meneim* "to be stable" and *monas* "oneness". The Circle is regarded as the ultimate unit of Being, which is essentially also all Being. In Blake's *The Ancient of Days*, God is represented as a geomancer and builder of the Universe. Symbolically He opens a pair of compasses to draw the Circle. The compass arms represents light rays shining from heaven toward the edge of the Circle. The Circle then encompasses a divine area of light.

Above **Compasses are instruments for drawing Circles.**

Right ***The Ancient of Days*** **by William Blake shows God using a pair of compasses.**

Above **This pool of water is a symbolic representation of a Circular container full of potential.**

Dimensions of Time

Imagine a single drop of water falling into the middle of a circular pool of water. Ripples are sent out as concentric Circles around their still Centre of origination. This pattern of rings around a Point is known as a Holon, which represents the dynamics of time. Each sequential Circle moves around the Centre just as every measure of a cycle in time – a second, an hour, a year and an eon – moves around the Moment. The indivisible Moment supports the ultimate Circle of eternity and the vessel of the Void. A Holon is therefore the structure of the Grid of Time, where increasingly long time cycles radiate outward like ripples from a Central source supporting it. This Centre is the indivisible Moment, the now, in all possible time.

Circular relationships, such as family circles or social circles, are bound in time. In these Circles we participate with others in events in time, yet each of us experiences these events from our unique perspective. Each participant also makes a unique contribution to the Circle. Communing is an intimate extension of our communication in our circular relationships as it includes a link between body and spirit on a personal level or with others in a community. Commune,

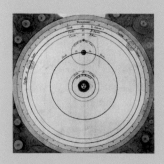

Above **In astrology this is the symbol for the Sun. The infinitely small Centre, the Point and Moment, supports the infinitely large Circle of the eternal.**

Below **French Copernican chart (1761) of a heliocentric solar system.**

The Circle and the Centre

In these examples Circles move around or contain a clearly defined, still Centre. In doing so they draw attention to it and signify its importance.

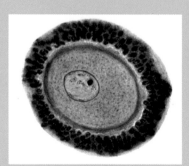

Above **The still eye of the storm (NASA).**

Above left **Human ovum.**

Left **Circles of refracted light around the central, radiant Sun.**

communicate, contribute, connect, community and continual cycles are examples of English words that all imply binding relationships that are inherently circular rather than linear. Continual change characterizes the passage and evolution of life; a shared experience involving an exchange of energy that is facilitated by the passage of circular time. Every moment in time we share events, not just with other humans, but also with everything else surrounding us. Every event, no matter how trivial, has a cause and effect.

As life moves in constant cycles, events form a complex web of links in an interdependent scheme of immense proportions within the Void. Events in time facilitate learning and ultimately contribute to the process of evolution on every scale. This Universe is just one turn of perpetual time as it revolves through successive cycles of creation and destruction.

Above **A galaxy rotating around its Central Point.**

"I shall now recall to mind that the motion of the heavenly bodies is circular, since the motion appropriate to a sphere is rotation in a circle."

NICOLAUS COPERNICUS,
POLISH SCIENTIST (1473–1543)

Holon images

The Grid of Time – a Circle filled with concentric Circles. Radial Circles around a Centre illustrate the expansive nature of consecutive cycles in time from the Moment.

Far left and left **Concentric Circles of the Holon, spaced out and close together, to demonstrate how they fill the Circle. The effects of events in the Moment radiate out to fill the Circle.**

Six pictures left **The Holon is evident in natural events, both great and small, as reality and metaphor. Circles form an impenetrable Holon – this can be observed in phenomena from solar system creation and layers of deposits in an agate crystal, to a spider's web and the rings seen in a halved onion.**

Circles of humanity

As a species, human beings form Circles to enhance the spirit of community and our intimate connection with each other. We use them to mark key events in our lives and create a sense of celebration in time.

Right and below **Maypole dancing marks May Day celebrations, while dance forms employ the Circle as a consolidation. The Circle is also a way to meet each other, communicate joy or anger and perhaps prepare for action, such as battle.**

Right **The world's time zones represented in a circular engraving in which numerous clocks are drawn, each representing a major world city. At the centre is a clock representing the time in Washington, DC, USA (by A. J. Johnson and Ward, 1862).**

God is an intelligible sphere, whose centre is everywhere and whose circumference is nowhere.

HERMES TRISMEGISTUS, LEGENDARY FOUNDER OF ALCHEMY AND ASTROLOGY

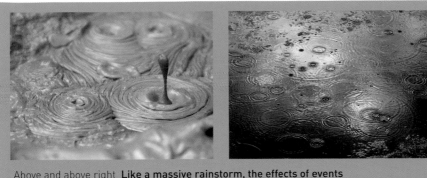

Above and above right **Like a massive rainstorm, the effects of events radiate out to fill the eternal, infinite Circle simultaneously.**

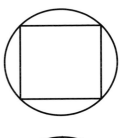

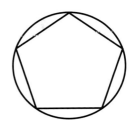

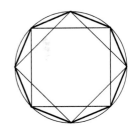

Unity

A true Point cannot be measured and is impossible to draw. Just like all the geometric shapes that come from it, Points are symbolic ideals. As a form of Being in its own right, the Central Point of a Circle is unity; a dot denoted by the number one, which can also be represented as an active, vertical 1. Like the Circle, everything created has a Centre within it, around which it revolves. This is the fundamental essence of the meaning of the word Universe – meaning "one turn".

1 is complete; it is whole and within it is every form of Being possible. Much like a fertile nucleus within an egg, number 1 contains the code for a new life to manifest into existence. So 1 is the ultimate fertile nucleus that has the latent potential to become anything conceivable, since everything conceivable is within it. A structured reality and multiple forms of Being do not fragment unity, unity remains, as does the Centre and the Moment. The Centre, the Moment, the 1 is everywhere as its ultimate expansion is the Circle that contains everything.

Scaling shapes up and down, supported by Circles

All geometric shapes and principles, no matter how complex, originate from the single Point at the Centre of the Circle. All have corners touching the boundary of the Circle that accommodates them.

Concentric Circles of the Holon establish the exact mathematical relationships necessary to construct the basic geometric shapes that recur throughout the Universe. Regardless of their size, the geometric relationships remain constant, no matter how large the concentric Circles become. Shapes structure reality on every scale of largeness and smallness; up to the infinitely large Circular vessel and down to the smallest Point. Logically the geometric code is therefore contained in the Point and permeates all reality.

Above **Examples of geometric shapes contained by, and supported by, the Circle.**

> The Being of God is like a wheel, wherein many wheels are made one in another, upwards, downwards, crossways, and yet continually turn all of them together. At which indeed, when man beholds the Wheel, he highly marvels.
>
> JAKOB BÖHME, GERMAN CHRISTIAN MYSTIC AND THEOLOGIAN (1575–1624)

Nothing then everything

0 1
  1 000000000...

When 0 has a single 1 added to it the 0 immediately gains value and now embodies the infinite possibilities that can be generated from 1, the Point. Adding an infinite number of 0000s is in effect the same as the expansive Circle, having the same shape as a 0, surrounding the Central Point, which is also 1.

Duality and perpetual rhythm

Imagine the Centre as a light source radiating out and bouncing off the distant horizon edge of the ultimately large, eternal Circle and returning to the Centre, generating its own reflection, just as we view our reflection in a mirror. So the Centre is active, looking and radiating, while the reflected image is passive.

Active and passive, this pair of opposites are to be called the Duals and they can only be defined in relation to the other, since neither is ever Absolute. Seamlessly they act in unison under tension. Such tension may be represented as a line, but no such line actually exists; there is only harmonious co-existence. With the Duals come discord and conflict, but also balance. Each reinforces the strength of its opposite. Each nourishes and supports the other through the perpetual rhythm of movement and change, as represented by the "S" in the T'ai Chi symbol. Tension of the opposites takes place in every natural occurrence and in daily events as contrasts and differences. All situations are in a constant state of change, with states in between such as extremes of freezing and boiling.

Boundaries and polarized views are illusions generated by duality and these deflect us from the inherent unity, the Central Point, within us. Concurrently duality repels and attracts, splits unity and yearns to return to it. This yearning is felt in our souls as a desire to return to the Centre and to the source of our existence. Duality, and hence tension and continual change, is within all subsequent shapes and every Being.

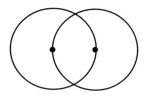

1. The original Point "sees" a reflection of itself on the boundary and a second Circle forms; each goes through the Centre of the other.

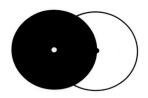

2. Imagine this as an active white Point surrounded by darkness and the reflection as its opposite, a receiving black Point surrounded by white.

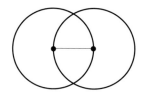

3. Tension exists between the two Points, like a force field, represented by a line. 1 + 1 = 2.

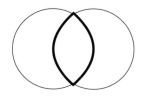

4. The outlined, overlapping area is known as the vesica piscis.

Duality

We see a reflection, an illusion of a face in a mirror. We also have symmetry in the face; the left and right sides are a reflection of the other, linked to the logical and creative sides of our vesica piscis-shaped brain (see picture below). This reflected duality is extended (see page 26) into the relationships between male and female.

Four pictures left **Mountains reflected in a lake, a face reflected in a mirror, two lovers and a butterfly with widespread wings all provide examples of duality.**

Below **The vesica piscis (see page 26) appears in numerous guises, in eyes, petals and many seeds. Note the brain, bottom row, second from left.**

Above **The mandorla (vesica piscis) is used in Christianity to depict sacred moments that transcend time and space, as shown in the ascension of Christ.**

Vesica piscis

The vesica piscis symbol is often regarded as a "door" for other numbers and is used as such in many church entrances. In the area where the Circles overlap the opposite principles blend. Visually and metaphorically it is like the opening (the female yoni) to the womb of creation.

The two Points where the Circles intersect then establishes the exact distances needed to construct the basic geometric shapes and patterns. Now we have the Circle, Point and a line with which shapes are constructed. The first three shapes to "emerge" are the Triangle, Square and pentagon.

Above **A Chinese example of the vesica piscis. Note the entwined dragons and Spirals.**

Above **A cathedral door arch based on the vesica piscis shape.**

Above **A vesica piscis in the middle of the Chalice Well pool, Glastonbury, UK.**

The T'ai Chi symbol

Taoist principles are summarized in the T'ai Chi symbol. Similar Western symbols are the half-black and half-white Circle that represent the source and union of the duality of the Universe, which act in a subtle balance. The S-shaped curve within the T'ai Chi shows how Yin and Yang support each other and the black and white dots show how nothing is totally Yin or Yang. According to Taoism, when there is perfect balance within yourself, it is possible to find perfect happiness in a state known as Wu Chi, where no duality or form exist, only wholeness. This is symbolized as the original empty Circle; a condition of emptiness waiting to be filled.

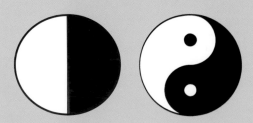

Far left **The half-white, half-black Western Circle.**

Left **The Yin-Yang of the T'ai Chi symbol.**

North and south

There is a continual tension between the Earth's magnetic poles and in the ever-cycling Duals of night and day provided by the Sun.

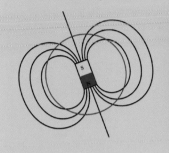

Far left **The Earth's magnetic poles.**

Left **Duals of dark and light dividing the Earth.**

> There must be a positive and negative in everything in the Universe in order to complete a circuit, or circle, without which there would be no activity, no motion.
>
> SIR JOHN A. MACDONALD, FIRST CANADIAN PRIME MINISTER (1815–1891)

Hydrogen

Hydrogen is the first atom of all the 92 to emerge and the most abundant in the Universe. It consists of only 2 parts, a positively charged proton at its centre and negatively charged electron orbiting it. All other atoms have a third neutron component and all are generated by fusing hydrogen atoms together within the extreme temperatures of stars.

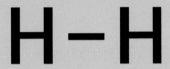

Left **The chemical structure of hydrogen. Note the dualistic style of the letter H as two lines joined together.**

1 and 2 are the "parents" of subsequent numbers

0	Absolute
1	Unity
2 = 1 + 1	Duals

- 1 is the only value that, when added to itself, produces a result greater than when it is multiplied by itself.
- 1 x 1 x 1 x 1 x 1 x 1...= 1
- 1 + 1 = 2, 1 + 1 + 1 = 3, 1 + 1 + 1 + 1 = 4 etc.

Sumerian words for "one" and "two" are also those for "man" and "woman". Representing the singularity and duality, 1 and 2 were not considered numbers themselves, but as "parents" of numbers that "mate" and "give birth" to all subsequent number principles. The innate human ability to give names to entities beyond number 2 is a major imaginative conscious skill that we take for granted.

3 Square, Space-Time and the Cross

Travel in all the four quarters of the earth, yet you will find nothing anywhere. Whatever there is, is only here.

RAMAKRISHNA,
19TH-CENTURY INDIAN MYSTIC
(1836–1886)

Above **The Square – foundations on Earth.**

As NUMBER 2 DOUBLES (2, 4, 8, 16, 32...), the Grid of Space forms, within which we can see the expanse of nested Squares (see page 29) that are a visual expression of the two dimensions constituting a surface or a plane, and the "four-square" foundations of our physical existence. Because of this the Square is often used to represent the surface of the Earth and it is used extensively for defining areas for the purpose of specific activities, such as rooms and town squares. In our story of geometry the Square provides a landscape, or stage, in the Absolute Void, like a map, filled with possible directions in which to move and experience events. Each location on the map of our lives is a setting on the stage where we act out our roles.

From our Point in space we have the option to move to any infinite number of other Points we perceive as "there". These are the Directions; the push and pulls of Duals evident in the opposing poles of north and south and in every decision we make. The Centre of the Square is Here; wherever we go we are always Here. Space appears vast externally, yet we only inhabit "Here". As Here is anywhere, it has infinite space within it.

Right **Squares and cubes are very rare in Nature. As a shape Squares are fragile and prone to having their corners knocked off. These examples show a pattern on a snake, pyrite and salt crystals.**

The Square in human design

Humans use the Square and cube prolifically in design and structures. Simple to set up and lay out, Squares are used in modern storage systems, building interiors, exterior layouts and road plans.

Above **This 18th-century Italian painting shows a tournament taking place in the city square.**

Three pictures left **Squares and cubes are evident in aspects of modern design and construction: in city skyscrapers, warehouse stacks and household filing systems.**

Spatial expansion in forms of Being

Squares double and expand to create the optical impression of space and expanse.

Right **Squares give a sense of position, a reference Point and distance. They provide perspective.**

Above **The crossroads literally represent the choices we make in life.**

Unfolding the Grid of Space

2 doubles, squares and expands infinitely.

Doubling	Squaring
2 + 2 = 4	2 x 2 = 4
4 + 4 = 8	3 x 3 = 9
8 + 8 = 16 ...	4 x 4 = 16 ...

2 is the only case where the addition to itself yields the same result as it does multiplying by itself.

2 + 2 = 4   = 2 x 2   2 doubled and squared

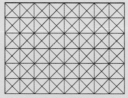

The Grid of Space

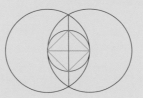

A Square within the vesica piscis

Above **When we are on the Earth we sense the Square grid that grounds us.**

Left **Sitting man "here" in a Square with the Directions to each side, as shown in this volcanic stone medallion, 14th–15th century, Dagestan.**

The 8 Directions

Opposing directions are indicated by the arms of a Cross in the middle of the main Square orientated toward the 4 Cardinal Points of north, south, east and west.

The Square may again be bisected by diagonals to divide it into 8 Triangles.

8 is significant as it is 2 cubed: 2 x 2 x 2 = 8. These diagonals are believed to point to the Four Corners of the Earth. Further division of the Square gives us 16 directions; 16 being significant in that 16 = 2 x 2 x 2 x 2, 2 to the power of 4.

Every time the Square is divided up in this way the exact centre of the Square is emphasized. This Centre is a unique location, or Point, in Space-Time, from which choices in direction can be weighed up.

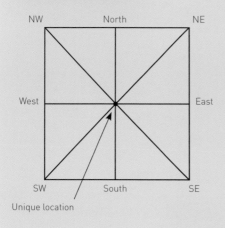

> I have not kept my square, but that to come./
>
> Shall all be done by the rule.
>
> WILLIAM SHAKESPEARE,
> ENGLISH PLAYWRIGHT
> (1564–1616)

Combining the Grid of Space and Time

As the Grid of Space unfolds, each step is always contained by a Circle of the Grid of Time. The simple symbol of a Square contained by a Circle summarizes the combined principles of the Grid of Time and the Grid of Space.

The Grid of Time combined with the Grid of Space

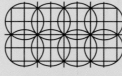

Circles and Squares combined

Square, Circle and Cross denote the Centre of space-time, Here, Now

The Square is implied by the Cross within a Circle

Union of Space-Time

Together the Square and Circle as part of the Grid of Space and the Grid of Time act in unison to provide Space-Time. Without space we could not shift position and without time there could be no motion. Both time and space act as one to facilitate our conscious experience of events in our own reality. Movement describes the dynamics of Space-Time, facilitating our life paths, or journeys, to

Squaring of the Circle

Squaring the Circle is a geometric principle known for many centuries and it is employed in sacred sites such as churches. When the area enclosed within their shapes, or their perimeter, is equal this is known as the Squaring of the Circle. Symbolically heaven and Earth, matter and spirit, are combined.

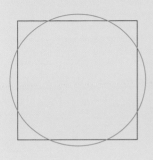

Far left **Circle and Square with the same perimeter.**

Left **Michael Maier's** *Atalanta fugiens, Emblema XXI* (1617).

learn, grow and evolve. This is the fundamental nature and purpose of life. Like a journey from home to work, we have to plan a route from beginning to end. We make choices about the direction we take and the Circles we belong to in every moment. Combined, the Circle and Square remind us that we are "Here, Now".

But why?

The Cross, with arms of equal length, is formed by the Square as it unfolds and this highlights the Centre. Radial Circles of the Holon draw further attention to the Centre. In spiritual terms it is "Here, Now" at the Centre, where your inner, immortal Self is waiting to be revealed by sacrificing your ego and its incessant demands (see page 81). On a Square, the horizons of our vision is limited so it is appropriate that our experience on the Earth's surface is represented as a Square. All Beings are eternal in Nature and that includes us, too, since everything comes from, and returns to, pure consciousness or energy; all Being, according to many beliefs. Yet on Earth we are mortal and subject to Destiny. The Circle is a reminder of this. The Circle links our choices into the ever-turning cycles and tells us that all our decisions affect everything else contained by the One Circle and vice versa.

As a symbol the Cross acts as a signpost that reminds us that we make our own choices in Space-Time as to where we go and what we choose to do. Since the Cross is integral to the Square and Circle it highlights the way in which every choice, in every moment, is linked into the Universal Plan. When making our choices the constant pull of opposites felt by our ego and emotions has to be recognized and resolved. So the Cross pertains to the question, "Why are we Here, Now?" We are Here, Now, to live and make balanced choices in every moment and to learn from them.

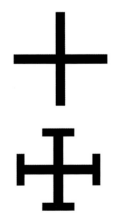

Top **Plain Cross.**

Above **Cross with ends, indicating that they are part of the Square and to show us the "doors" into Space, to the Centre.**

Above **French and English Cross medals used for rewarding choices of personal sacrifice.** Sometimes the choice involves the loss of a person's life.

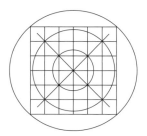

Square unfolds and Circle Holon radiates out

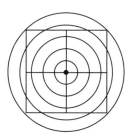

Holon and Cross highlight the Centre of Space-Time

Squares on the surface of the Earth curve when you view them from a distance.

The sacrificial Cross
The Cross represents our choices in life and the lessons we learn. It is a powerful symbol found in religion, art and geometry.

Far right **19th-century depiction, by Gustave Doré, of Christ on the cross.**

Right *St John Chrysostom* (c347–407) the Golden Tongued, Bishop of Constantinople by Pedro Orrente, 17th century.

Doors between material and spiritual worlds
Hindus believe that at the end of each arm of the Cross are 4 doors, gates or portals between the material and spiritual worlds. Metaphorically, the Cross implies that the worst thing that happened to you while on Earth could become the best. A symbolic union of the Cross and the Sun sign, a Circle with a dot in the middle, implies that through our physical manifestation on Earth, finding our Inner Self requires us to look at ourselves in a mirror to find our own inner light, or Sun.

Pre-Columbian Sun symbol

The Celtic Cross enclosed in a Circle

The power of intent
Intent is the motivation behind everything we think, say and do and then this intent underpins the impact of an event (be it a thought, a word or a deed), the effects of which instantaneously radiate from the Central Point "Here, Now" to affect everything else in the Circle.

Intent can be sensed: it can be felt in the energy in another person's eyes as they look at you across the room, even when you are not looking at them; it can be experienced as healing energy from another person's hands; it is known when someone says one thing yet means exactly the opposite; it can even be anticipated; it is in our Hearts when we weigh up our choices right Here in the Moment. Is our intent bad or good? Will it make us rich or poor? How will it make us or others feel? What are the possible outcomes? Our choices on our Cross are imbued with our intent from the instant we have a thought, and our thoughts create our reality.

Wheels and spokes

Division of the Circle into 4 is used as a Sun symbol and also as an astrological representation of the Earth. Division of the Circle into 8 is symbolic of the 8 Directions in the Wheel of Life. Humans have divided the Circle into 4, 8, 16, 32... uniting time and space. Nature similarly divides the Circle into numerous spokes and sectors.

Above **Doors lead to an exceptionally long corridor in this Hindu temple, which is lined with sculptured and painted pillars. Wooden double doors are at the far end. We feel as though we are entering the Centre.**

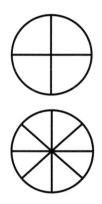

Above **The Circle is divided into 4 and 8.**

Our intent revealed

Division of the Circle into 4 also represented the 4 chambers of the human heart, which is where we weigh up our choices emotionally and feel the consequences. Intent starts in the Mind, comes from the heart and can be seen in the eyes.

Four pictures left
As an ancient symbol, the Circled Cross mirrors the chambers of our physical heart. See and feel the intent in the eyes of this cat and man, intent that has originated from their hearts.

33

4 Triangles and Being

All things in the world come into existence from being, and being comes from non-being.

LAO-TZU, CHINESE PHILOSOPHER
(C6TH CENTURY BC)

Above **The Triangle of creation, strength and stability.**

TRIANGLES ARE CENTRAL TO THE ACT OF CREATION, THE PROCESS OF CREATION and the essence of Being. The Triangle represents the Trinity, comprising opposites and their union to create the third. As opposed to swinging from extremes, the third stabilizes, strengthens and transforms the Duals by raising the sum of their parts to a higher level. It provides a middle option, for example up/middle/down, hot/warm/cold. Without the possibility of this position of stability, the opposites could not be viewed and weighed up for their contribution in a decision. This is why scales are used to represent justice and in Egyptian mythology symbolically the Heart is weighed (see box below) to determine the worth of the individual based on their life choices.

Act of creation

Number *3* is a sacred number in most religions as it combines the number *1*, the first, and the *2* of the Duals: 1 + 2 = 3. So number 3 is the act of creation and believed to include all life and experience. Numerous groups of 3 are found in all scientific and metaphysical studies of the structure or organization of physical life. The evidence highlighting the role of number 3 in the fabric of life highlights its pivotal role in completely defining what it is "to be".

Above **Cards balanced in Triangular formation.**

Weighing the soul
Scales represent justice. Symbolically, hearts are weighed to see how worthy the soul is.

Far right *Last Judgment*, 13th-century chromolithograph from a psalter.

Right **An Ancient Egyptian funerary papyrus.**

The Rhombus or Diamond

Where the Dual Circles cross, their place of intersection creates a third Point and establishes a Triangle contained within the vesica piscis.

Union of unity and duality
2 + 1 = 3

Two equilateral Triangles with a common base make a rhombus. From the "father" (male) and "mother" (female) principles a "son" and "daughter" are born at the top and bottom, where the Circles cross. A "diamond"-shaped rhombus is now within the vesica piscis.

The Triangles' common base is the line between the Duals, delineating "above" and "below". In many belief systems the upper Triangle is the spiritual realm; the lower Triangle the physical. These realms cannot exist without each other, as a mirror's reflection cannot exist without a solid form.

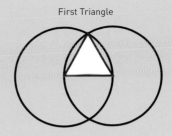

First Triangle

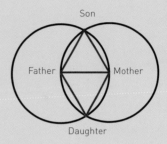

Rhombus or diamond

Son

Father   Mother

Daughter

Triangles in religious art

Trinities feature in most belief systems, often with Triangular symbolism. Below are examples from a selection of cultures.

Above left **Brahman with 3 heads (Hindu).**

Above **Triple-headed God Apedemak, "the Lord of royal power" (Ancient Nubian belief).**

Above right **The Trinity of Osiris, Isis and Horus (Ancient Egyptian belief).**

Left **Symbol enclosing a Triangle, surrounded by a Circle with 8 trumpets (Christian).**

A tragedy is a representation of an action that is whole and complete and of a certain magnitude. A whole is what has a beginning and middle and end.

ARISTOTLE,
GREEK PHILOSOPHER
(384–322 BC)

Above **Triangle within the Circle of time.**

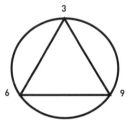

Above **Asian Indians believe that numbers 3, 6 and 9 (themselves a Trinity) tie the creative process and the cycles of cosmic time together.**

Above **Trefoil – representation of the Christian Trinity, comprising 3 intertwined vesicae piscis.**

Above **Examples of Triangles in Nature.**

Raising by reconciling

By reconciling their differences, the sum of the Duals' parts is raised and transformed to a higher level, just as male and female combine to fertilize an egg that has extended the evolutionary process and maintained the process of change. In the Celtic world the number 3 was deeply auspicious. When present or repeated, it had the effect of strengthening and intensifying energy and form.

Fertile number 6

From a spiritual perspective the union of the Duals, of male and female, of matter and spirit, fertilizes the ocean of unmanifest consciousness pervading the Void. By vibrating at different levels of frequency various forms of Dual Beings of matter/spirit, male/female can become manifest and exist in this energy ocean of infinite possibilities. Such forms of Being exist at different vibrational levels.

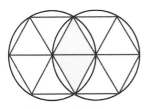

Rhombus within the vesica piscis

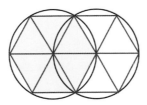

Hexagon within a Circle
3 + 3 = 6

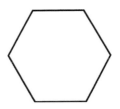

Hexagon

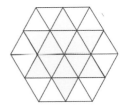

The union of Duals in perfect balance with the hexagram, or Star of David, within the hexagon
1 + 2 + 3 = 6   3 + 3 = 6

Star of David

The House of David is represented by the hexagram, or a star with 6 points, comprising 2 Triangles intertwined; the downward Triangle of Spirit that intersects with the upward Triangle of Matter, having a common Centre. The "seminal ones" were the Semites, descended from David. Hence the hexagonal star is also known as the "Star of David".

Above **Late 16th to early 17th-century Star of David and 6 Circles within a Circle. (From a manuscript of the Hebrew Bible.)**

Above **The Leningrad codex, a former possession of Karaïte Jews, c1010. Note the 2 Squares, indicating the 8 Directions, all lying within the Circle.**

Cosmic womb

Vishnu, God of Creation, has a mark made of 2 superimposed inverted and upright Triangles that also represent Siva and Sakti. Unity of these Triangles, emphasized by the Circle around them, is the foundation of duality. Surrounding them is the "primal Triangle" representing the "cosmic womb" of the creative female Sakti.

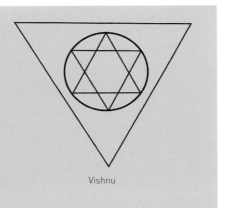

Vishnu

> Nothing can
> be created
> from nothing.
>
> LUCRETIUS, ROMAN
> POET AND
> PHILOSOPHER
> (c99–c55 BC)

Hexagons

The hexagonal shape occurs naturally throughout Nature and is especially prevalent in crystals, mostly because 3- or 6-sided meshes and structures are strong and stable, just as our own human applications of this shape in fishing nets and tripods. Honeycomb comprises 6-sided cells and many philosophers viewed the honeycomb as the manifestation of divine harmony in Nature.

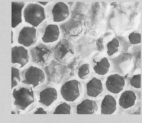

From microscopic snow crystals to magnificent basalt rock formations, hexagons are to be found everywhere in the natural world.

Above **Number 9 is represented as an expanded rhombus in many areas of mysticism. Note that it includes a hexagon within its boundary.**

Above **Rhombus-shaped ray.**

Perfect number 9

As the last number before number *10*, where *10* is in itself a visual form of the Void (0) and originating single Point (1), number *9* is the number of the Perfect Man, who has mastered the lessons of life as the soul passes through *9* stages. Reflecting the *9* Planes of Being, *9* is the ultimate number in the sequence from *0*, *1* through to *10*, to be reborn as an enlightened Being.

The Ancient Greeks referred to the number *9* as the "horizon" (the Ennead), which lay at the edge of the endless ocean of numbers. These repeat in cycle since the digits of any number multiplied by *9* always total *9* when their digits are added. Ancient Hebrews thus believed that *9* was the number of immutable Truth, since it always produced itself and returned to itself while encompassing all the numbers within it.

3 + 3 + 3 = 9   3 x 3 = 9   9 perfection and a Trinity of 3
9 = 3 squared (3 x 3) = 1 cubed (1 x 1 x 1) + 2 cubed (2 x 2 x 2)
9 represents the principles of 3 taken to their greatest expression

Examples:
7 x 9 = 63 and 6 + 3 = 9
12 x 9 = 108 and 1 + 0 + 8 = 9
40 x 9 = 360 and 3 + 6 + 0 = 9

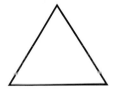

3: Union of "parents" and creation

The Enneagram
A figure with 9 points, called an "enneagram", may be drawn using a continuous line, where every alternate point is joined.

Right **The popular Enneagram of Personality and the Fourth Way teachings use an irregular enneagram consisting of a Triangle and an irregular hexagram.**

Far right **Example of an enneagram.**

6: Fertility and duality within Being

Planes of Being

At the core of many beliefs is the concept of 9 hierarchical Planes (see box below), or levels of Being, comprising different subtle vibrations. These Planes span from the slowest material plane to higher, finer and more ethereal Planes and mirror the progression of the human soul as it evolves into an enlightened Being. This is reflected in the material world as the 9-fold rule. For example, in medieval Europe there were 9 types of crowns in heraldry. The founders of the United States government used the Egyptian Ennead as their model for the 9 justices of the Supreme Court.

9: Perfection and creation in progress. Star figure made of 3 Triangles

Levels of Being
Aztec, Mayan and Native American myths refer to 9 cosmic levels – 4 above, 4 below and 1 in the middle. Taoists believe in 9 upper levels, leading to heaven and 9 lower levels, leading to the abode of the dead. The "Celestial Hierarchy" of Christianity, the 9 Circles of heaven and 9 choirs known as the "heavenly host" surround the throne of God. Only the two lowest categories of the heavenly host in direct contact with Man announce God's will, protect the virtuous and punish sinners. In opposition there are 9 orders of devils within 9 Rings of Hell.

Right **Note the Triangle circled by 9 at the top of this Hierarchy of Angels.**

Above **Dante navigated the 9 Rings of Hell and 9 Spheres of Heaven before his journey culminated with his ascent to the tenth realm of transcendent spiritual splendour. Dante and Beatrice gaze upon the highest heaven in *The Empyrean* by Gustave Doré (1832–1883).**

Replication of 3 in the Grid of Being

The Grid of Being replicates with the number 3 as its basis and is structured around equilateral Triangles. Like the Squared Grid of Space and the Circular Grid of Time, the Grid of Being is a basis on which the nature of Being can be analysed, understood and represented. Duality is still evident as the aspects of spirit and matter, heaven and Earth.

In Indian systems 3, 6 and 9 are associated with the creative process, Cosmic Time and its cycles. The numbers 3, 6, and 9 unite as a further Trinity, embodying the circling cycle of creation and birth from 3 (the first step of creation), to 3 + 3 = 6 representing fertility and the union of male/female, matter/spirit in the realm of heaven and Earth, and finally completion represented by the eternal 9, where 9 = 3 + 3 + 3: another Trinity.

The Tetraktys, or Decad

The Third-Level Triangle is a very significant triangular model called the Tetraktys, which is Greek for "fourfold". 9 small Triangles are contained within a larger tenth Triangle. So revered was this ancient symbol that it inspired ancient philosophers to swear by the name of the person who brought this gift to humanity, Pythagoras.

The Tetraktys appears regularly in future chapters (see pages 61, 91, 121 and 128). In it there are 9 smaller Triangles to reveal the larger tenth Triangle and number 10, being made of the 0 and 1; symbolically it represents the individual within the infinite Circle. Another significant geometric shape contained by the Tetraktys is the 6-sided hexagon.

The 3 Steps of creation

Every number between 1 and 10 can be produced as a sum of the numbers 0, 1, 2, 3 and 4 that comprise the Tetraktys. So by taking "3 steps" the foundations of reality are established. From 10 onward the numbers cycle, hence 10 is the number of Completion.

0 – Circle	1
1 – Unity	1+1 = 2
2 – Duals	1 +2 = 3
3 – Triangle	1 + 3 = 4
4 – Square	1 + 4 = 5 = 2 + 3
	1 + 2 + 3 = 6 = 2 + 4
	3 + 4 = 7
	1 + 3 + 4 = 8
	2 + 3 + 4 = 9
	1 + 2 + 3 + 4 = 10

Significant Levels of Triangle within the Grid of Being

1 + 2 = 3   First-level Triangle
1 + 2 + 3 = 6 = 3 + 3   Second-level Triangle
1 + 2 + 3 + 4 = 10   Third-level Triangle

The second-level Triangle unfolds itself to form the Grid of Being using the number series produced by the addition or multiples of 3.

The digits of each number created in the 3 number series totals up to 3, 6 or 9 in a repeating cycle, 3, 6, 9, 12 (1 + 2 = 3), 15 (1 + 5 = 6), 18 (1 + 8 = 9), 21 (2 + 1 = 3) and so on.

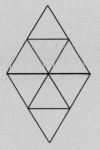

As the Triangle replicates, so do the Duals.

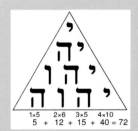

First-level Triangle
3 points
1 + 2 = 3 Creation

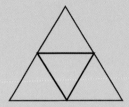

Second-level Triangle
6 points
1 + 2 + 3 = 6 Fertility

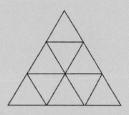

Third-level Triangle – Tetraktys
10 points
1 + 2 + 3 + 4 = 10 Completion

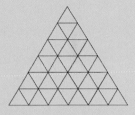

Grid of Being

"The Ennead flows around the other numbers within the Decad like an ocean."

NICHOMACHUS, MATHEMATICIAN (AD c60–c120)

The Tetragrammaton

Hebrew letters of the Tetragrammaton arranged as a Tetraktys, showing that by the rules of Geomatria (giving numerical values to words or phrases) the sum is 72 (see also page 143). "Tetragrammaton" refers to the Hebrew name of the God of Israel, YHWH, used in the Hebrew Bible.

Above **Tetraktys (from a diagram by German Hebraist/Kabbalist Johannes Reuchlin).**

Left **12th-century manuscript of c1109. The 3 parts of the Christian Trinity are assigned to the 3 possible consecutive 2-letter sequences of the Tetragrammaton (IE, EV, and VE). These 3 sequences are in 3 Circle nodes, connected to each other as a Triangle. In the Centre is the complete Tetragrammaton.**

41

5 Matrix of Space-Time-Being

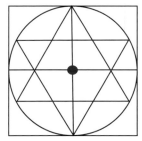

Above **Matrix of Space-Time-Being.**

I know that the study of material things, number, order and position are the threefold clue to exact knowledge; that these three, in the mathematicians' hands furnish the first outline for a sketch of the Universe. … For the harmony of the world is made manifest in form and number, and the heart and soul and all poetry of Natural Philosophy are embedded in the concept of mathematical beauty.

SIR D'ARCY WENTWORTH THOMPSON,
BIOLOGIST, MATHEMATICIAN, CLASSICS SCHOLAR (1860–1948)

> The web of our life is of a mingled yarn, good and ill together.
>
> WILLIAM SHAKESPEARE, ENGLISH PLAYWRIGHT (1564–1616)

WE HAVE REACHED THE STAGE IN OUR JOURNEY THROUGH GEOMETRY where a superbly simple geometric model and symbol, comprising a combination of the Circle, Square and 2 superimposed equilateral Triangles, can be put together. Its two-dimensional surface is the essence of simplicity. However this model embodies underlying complexity. This unified symbol is called the Matrix of Space-Time-Being and it establishes the basic framework of the Gateway to the Heavens model (see page 197). Within this Matrix is an implied Cross, highlighting the common Centre of the 3 primary shapes, the Circle, Square and Triangle, which correlate with the 3 essential aspects of time, space and Being.

"Humankind has not woven the web of life. We are but one thread within it. Whatever we do to the web, we do to ourselves. All things are bound together. All things connect."

CHIEF SEATTLE (1780–1866)

"We sleep, but the loom of life never stops, and the pattern which was weaving when the sun went down is weaving when it comes up in the morning."

HENRY WARD BEECHER, CLERGYMAN, SOCIAL REFORMER AND ABOLITIONIST (1813–1887)

First stage of building up the Gateway to Becoming

Each of the grids may be considered in isolation as well as how they combine to make the Matrix of Space-Time-Being. Within the Matrix are the key basic geometric shapes and their principles, imbued with the dynamics of duality.

Grid of Time

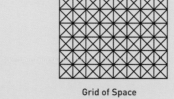

Grid of Space

Grid of Being

Circle
0

Point
1

Duality
2

Square
4, 8

Cross

Triangle
3

Hexagram
6

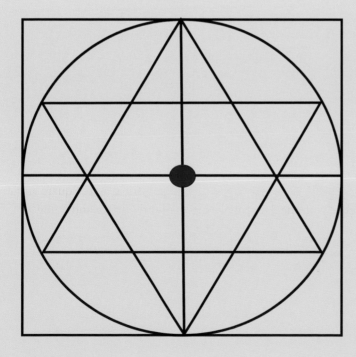

Left **Simple representation of the Matrix of Space-Time-Being, combining the essence of the Grids of Time, Space and Being, where the Cross highlights the Centre and 4 doors to the 4 Directions (see page 30).**

ESSENTIAL FEATURES OF THE MATRIX

Absolute essence	Aspect	Shape		Central location
Eternity	Time	Circle	●	Moment
Infinity	Space	Square	■	Point
Sentience/Existence	Being	Triangle	▲	Eye

MEANINGS OF THE CIRCLE, SQUARE AND TRIANGLE

Summarizing their associated descriptive English words and their first letters, which are similar to the shapes in appearance.

	Circle	Square	Triangle
Prime letters	C	D	V
Structural words	Communicate	Domain	Vibration
	Commune	Distance	Vessel
	Connect	Direction	Vision/View
	Community	Domicile	Variety
	Continual		
	Change		
Active words	Evolution	Movement	Energy
Outcomes	Events	Motion	Existence
	Experience		Ethereal

Summary of the first five chapters

Each of the simple, basic geometric shapes has a critical role to play in holding and sustaining reality. Not only do shapes structure reality, they also facilitate dynamics, such as communication and a sense of belonging, without which our lives would be meaningless.

GRID OF TIME

Circular relationships turn in successive cycles of continual change. All are bound in eternal time that is just one turn of the perpetual "wheel" containing every possible event in time. This is a larger version of *samsara*, the cycle of rebirth in which individual souls are reincarnated. Connections made through interactions are a shared experience, a communion between all Beings at a physical and soul level. A highly complex web of interconnecting events within the Moment are part of a scheme of immense proportions that is the evolutionary process.

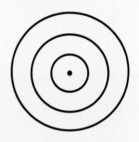

GRID OF SPACE

Squares of space ground us and provide us with a stage on which to act out our lives. From our Point Here in space we choose from the opposing directions and as such influence the journey of our self-development along the Spiritual Path. Within the Square is the Cross, reminding us how every choice counts, Here and Now.

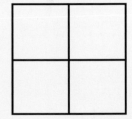

GRID OF BEING

Triangles enable creation, fertilization and perfection of Being. Through their union the Duals fertilize the Void of infinite possibilities and the third is born, setting creation in motion as rippling vibrations. Different degrees of vibration permeate the planes manifesting Beings of matter and spirit. No Being is ever static, but continually shifts from one condition to another.

THE CROSS

In every Moment we make choices that are part of what spiritual theorists term the Universal Plan. Underlying choice is the Cross, imbued with the forces of Duality. Duality within space, time and Being is experienced as the constant pull of opposites felt by our ego and emotions, and it is these we have to resolve when making our choices along our journey. We are Here to "make choices" in recognition that as all forms of Being are interconnected with all other Beings our choices affect everything in the Moment.

6 Spiral Life Force

> Progress has not followed a straight ascending line, but a spiral with rhythms of progress and retrogression, of evolution and dissolution.
>
> JOHANN WOLFGANG VON GOETHE,
> GERMAN WRITER AND POLYMATH (1749–1832)

Above **The Wheel of Life is set in motion by the Life Force, represented by water.**

Below **Spirals as they occur in Nature: in animals and plants.**

CIRCLES ESTABLISH TIME, SQUARES DESIGN SPACE AND TRIANGLES are the basis of Being. This Trinity of simple, yet profoundly important, geometric shapes together generate an impenetrable mesh; a Matrix of Space-Time-Being permeating reality on every scale imaginable. But what is the force behind reality that provides the impetus and momentum that stimulates and moves it? What animates, sustains and enables growth and expansion, contraction and decay in creation? The answer is the Life Force and the Spiral is the ultimate expression of the Life Force in action on every scale, whether in galaxies, water, air or the humble shell.

Above **Spirals occurring in the Elements of earth, water and air.**

Above **A pair of interacting galaxies named ARP 272 (Hubble, NASA).**

Above **Clouds off the Chilean coast show a pattern called a "von Kármán vortex street", which has been vital in the understanding of turbulent fluid flow that underlies a wide variety of phenomena, such as the lift under an aircraft wing.**

Spiral labyrinths

Labyrinths take the form of a Spiral and can be seen as metaphors for our meanderings on Earth, as we try to discover our inner Selves. The labyrinth's path takes you to its Centre and returns you to its entrance.

Hindu or Indian form of Spiral labyrinth

Cretan labyrinth

Intricate European labyrinth drawing

> The human mind always makes progress,
> but it is a progress in spirals.
>
> ANNE LOUISE GERMAINE DE STAËL-HOLSTEIN,
> FRENCH WRITER (1766–1817)

Spirals in human creation

As in Nature, Spirals are used extensively by humans, featuring world-wide in petroglyphs and extensively in sacred art and architecture.

Spiral staircase (Spanish)

Wood-carved door lintels (Maori)

Church spire (Danish)

Carved megalithic stone (Irish)

Gold plate (Celtic)

Spiral minaret (Iraqi)

> All rising to great places is by a winding stair.
>
> FRANCIS BACON, ENGLISH PHILOSOPHER, STATESMAN AND WRITER (1561–1626)

Serpents and the Life Force

Fertility and generative power are almost universally symbolized by the serpent coiled into a Spiral formation. The serpent which swallows its own tail (the Ouroboros, see left) is symbolic of the eternal cycle of the Circle.

Far right A snake coiled around the branch of a tree.

Right Coiled snake sculpture at the north altar of the Tenayuca pyramid in Tlanepantla, Mexico.

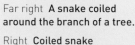

Naming and symbolizing the Life Force

Chinese	*chi*	Pacific Islanders	*mana*
Japanese	*ki*	North Africans	*baraka*
Hindus	*prana*	Greeks	*pneuma*
Kabbalah devotees	*nefish*		

Other names include *po-wa-ha*, *kurunba* and the *orgone* of Wilhelm Reich.

There are numerous examples of spiralling representations of the Life Force, often within a Circle. Examples exist on petroglyphs, tombs, temples, as body art, in representations of gods, on pottery and other artefacts, some dating to the Stone Age. The entrances to megalithic burial chambers in many parts of Europe are covered with Spiral patterns, assumed to denote the journey of the soul (see photograph of the carved entrance stone at Newgrange, Ireland, facing page).

Above **This fern beautifully illustrates the unfurling of Spirals within Spirals.**

Back and forth

Tracing a path from, and back to, the Centre, the Spiral follows a path along which evolution and involution, expansion and contraction can move. From the Centre the Life Force radiates out, sustaining reality and Being, and returns to the source. Within the Trinity of each of the 3 primary shapes we see the Spiral at work, stimulating their propagation and sustaining them. The Matrix of Space-Time-Being is animated and sustained by the Spiral Life Force.

Spirals within Squares and Triangles

Spirals within geometric shapes, such as Squares and Triangles, allow the expansion and contraction of the shape's essence from the Centre, due to the Life Force activating them.

Three triskeles

Solar Cross (top) and pentaskelion (above)

Above **A blue glow of young stars outlines the Spiral arms of galaxy NGC 5584. Dark dust lanes stream from the yellow core, where older stars are. The reddish dots are mainly background galaxies (NASA).**

The expanding Universe and dark energy

NGC 5584 was one of eight galaxies astronomers studied to measure the Universe's expansion rate. Albert Einstein assumed that the Universe was static and he hypothesized that there was a pushing force at work called the "cosmological constant", which counterbalanced the pull of gravity so that the Universe would not collapse. Then, in 1923, Edwin Hubble found that galaxies were moving away from us at a rate proportional to gravity, called the Hubble Constant, which meant that the Universe was uniformly expanding.

More recently, while determining how this expansion was expected to slow down over time, two studies (one led by Adam Riess of the Space Telescope Science Institute and the Johns Hopkins University and Brian Schmidt of Mount Stromlo Observatory, and the other by Saul Perlmutter of Lawrence Berkeley National Laboratory) independently discovered "dark energy", which seems to behave as a steady push, accelerating expansion, just like Einstein's cosmological constant. Note that as galaxies grow and evolve they take on Spiral formations (see also page 154).

Sharing for mutual advantage

The generative Spiral is basic to pattern-forming and, as is shown later (see page 62), shared patterns are required for Nature's myriad creations. Sharing is a reciprocal process of cooperation. It is the Dualistic interaction of active giving and passive receiving that sustains life. Sharing is a condition of survival and a critical factor in evolution. The symbiotic nature of sharing emphasizes the mutual advantage gained from the act, with positive results for both giver and receiver. The results of sharing coil out within the concentric Circles of the Holon and expand the Life Force further. Ever-increasing arrangements of mutual evolution, the unfolding of the Triangle, result in a plethora of forms structured by geometry. Variety is the accomplishment of creativity and sharing facilitated by the Spiral, spreading out as ever-increasing arrangements of mutual evolution.

Turning and transforming

In an Archimedean Spiral the distance between each coil remains the same, as seen, for example, in a coil of rope, a clock spring and a toilet roll. A helix is a three-dimensional version of the Archimedean Spiral as found in a DNA molecule and in screws and bolts. Strictly speaking the helix is not a true Spiral as it does not expand. Unlike the expanding Spiral generated from sharing ratios, the helix turns and stirs, to transform Being to a higher level.

Coming from opposite polarities, the double helix shows us the generative power of these opposites due to the reciprocal interchange of energy through the act of sharing. This is seen in the union of male and female sperm and egg and their sharing of double helix DNA, resulting in a new life. The motion from one state to the other ensures that each perpetually and seamlessly sustains the other.

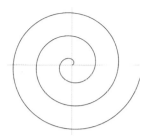

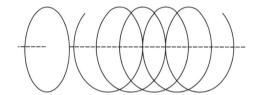

Left **From the "top" (far left of diagram) a helix appears as a rotating Circle around a Point: the Moment. From the side we can see it moving along, driven by the Life Force around an axis.**

Top and above **Archimedean Spiral, as a diagram and in a coiled fern.**

Above **Examples of the Archimedean Screw.**

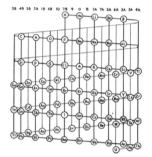

Above **Periodic Table of chemical elements depicted as a helix.**

The infinity symbol

A simple representation of the double helix is the modern infinity symbol, which originates from the double-ended Spiral. In essence the T'ai Chi symbol also shows us the blending of the Duals through the Spiral S-shape running through the black and white.

Above **The T'ai Chi symbol's constituent parts of opposite principles connected by the double-ended Spiral.**

Above **The modern infinity symbol and the symbol from which it originated, called the "lemniscate".**

Generative power of opposites

In the double helix we can see the vesica piscis, the shape created where the Duals overlap, repeated in every turn. Opposites blend in the vesica piscis and this gives the whole helix its generative capability (see also page 26). Often this is shown as 2 serpents coiled around a central staff, which is the still axis of the Centre.

Top **DNA (deoxyribonucleic acid) is the hereditary material in humans and almost all other organisms.**

Right to far right **Examples of the double helix: a pre-dynastic Egyptian gold knife; two versions of the Caduceus staff.**

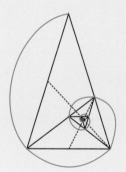
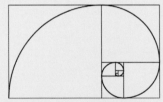

Above **The Phi Ratio Spiral in a rectangle.**

Left **The Golden Triangle making a Fibonacci Spiral.**

Phi in polygons

Regular polygons (geometric shapes with many angles and sides, usually more than four) can be increased while still maintaining the ratio of their sides. This ratio is variously known as the Phi Ratio, Sacred Cut, Golden Mean, Golden Ratio, the Divine Proportion and the Golden Section. This division of space is aesthetically pleasing and is used extensively in art and architecture.

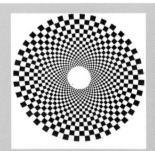

Above **The Phi Ratio's double Spiral tessellation displayed in a strawberry, sunflower seeds and pineapples.**

Ratios and generative Spirals

Expansive Spirals emerge through mathematical relationships, particularly the Phi Ratio. This is an irrational number appearing everywhere in Nature and discussed at great length in numerous texts. Phi cannot be expressed in terms of the ratio of two whole numbers; rather it is a number that has an infinite number of decimal places: Phi = Φ = 1.6180339…. It is better to think of Phi in terms of relationship rather than numbers and it can exist between 2 measurable quantities of any kind.

Irrational numbers, such as Phi, are best viewed in geometric form. When a geometric shape uses the Phi Ratio it is said to be "dynamic". In essence, the Phi Ratio, Spiral and pattern-sharing are united in animating and generating life.

Fibonacci Series
The expansive Fibonacci Series is the secret to self-replicating growth in geometry and, hence, Nature.

$0 + 1 = 1$ $1 + 1 = 2$ $1 + 2 = 3$ $2 + 3 = 5$ $3 + 5 = 8$ $5 + 8 = 13$

21, 34, 55, 89, 144, 233……is an accumulative process that grows from within the original 0 and 1; the Point within the Void.

Bees, hives and fertility
Unfertilized female bees give birth to males (or drones), but fertilized females always give birth to females. This pattern of genealogy maintains the balance of the collective hive. Hence the family tree of the beehive comprises the accumulative Fibonacci growth rhythm, while each branch resembles the whole family history.

Concentration of energy
The Life Force within a Triangle flows toward the apex, just as the water going down a plug hole coils into its conical apex and air in a tornado.

Right **Spirals from the "Heavens" to Earth; energy concentrating the closer it gets to the Centre.**

Above **A rose, showing its petals spiralling into the Centre.**

> Force never moves in a straight line, but always in a curve vast as the Universe and therefore eventually returns whence it issued forth, but upon a higher arc, for the Universe has progressed since it started.
>
> Kabbalah

7 Numbers 5 and 10

Above **5 was a key number for the Maya who considered it to be the fifth Direction at the Centre of the cardinal Directions, as shown here in their Sun calendar.**

It is a frequent assertion of ours that the whole Universe is manifestly completed and enclosed by the Decad, and seeded by the Monad, and it gains movement thanks to the Dyad and Life thanks to the Pentad.

IAMBLICHUS, ASSYRIAN NEOPLATONIST PHILOSOPHER (AD c245–325)

NUMBERS 5 AND 10 HAVE ENORMOUS SIGNIFICANCE in many cultures. This chapter explores how 5 and the sacred symbol of the pentagram relate to the perfection of form. It is a recurring theme that number 5 is the "limit" that gives bounded form to Being and introduces life itself via a perfect balance of active and receptive pairs infused by the Life Force.

Number 5, the Pentad

Number 5 is key to Nature's generative fertility, diversity and sharing. In essence it is all about versatility. As part of the process of creation, 5 underlies the composite structure of its creations, as well as the characteristics and personality that define their uniqueness. In the human we can see this in our 5 physical senses.

Above **The Aztec 5 Directions, with the deity for fire in the Centre, were developed into a ritual calendar of 260 days.**

Right **Pentagons and pentagrams in Nature: 5 is the most typical number in living Nature, especially in plants.**

There are two ways to look at the composition of 5 and each reveals its nature. The number 5 is the fifth Fibonacci number: 5 = 2 + 3, combining the Duals and Being. Also 5 = 4 + 1, where there are four balanced forces at work (two pairs of Duals) around the 1.

5-petalled flower

Pentagram and perfection of form

The regular pentagon is a uniform, 5-sided structure. By joining the apexes of the pentagon a star shape known as the "pentagram", or "pentacle", is outlined. Each of the Triangles in a pentagram has two equal sides that relate to the third side as a Golden Ratio and the Fibonacci Series and, because of this, the generation of the Spiral Life Force. The pentagram is a visual version of the Phi Ratio and Golden Mean. Effectively this star represents the generative Life Force, since the Spiral is in itself a concept, a pattern, not a shape.

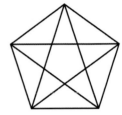

Pentagram (the 5-pointed star)
inside a pentagon

Pythagorean Triangle

The sides of 10 right-angled Triangles within the pentagon approximate to 3:4:5 unit lengths. This 3:4:5 Triangle is known as the Pythagorean Triangle, which is frequently found in plant patterns. Each larger (or smaller) section of the Triangle is related by Phi Ratio. When folded up as a 5-sided pyramid, imagine the peak transmitting out to infinity, like a huge aerial. In fact, the structures of aerials are very similar in design.

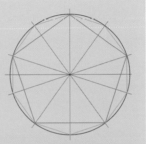

Above **10 Pythagorean Triangles, with one common corner at the Centre of the Circle.**

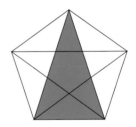

Golden Triangle in a pentagram

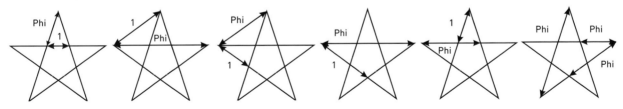

Pentagonal star and inherent generative Phi proportions

Old solar year

In ancient times a week comprised 5 days and the solar year was based on number 60 and could be divided into 5 x 72 days. This was later changed into a solar year of 365 days by adding 5 days, called the *epagomeneia*.

Above **You can see the rotation of the Spiral in the formation of this flower's petals.**

Top and above **As seen in this cluster of pencil points, pentagons cannot be perfectly aligned; reflecting the dynamic of 5 in that it mixes up order. But it can be used with other geometric shapes, such as hexagons in a ball.**

Venus, the Morning Star and Evening Star

The course of planet Venus can be plotted in the Zodiac when it appears as the Morning Star and the Evening Star. For 247 days Venus is visible as the Evening Star, representing Goddess Venus, symbolizing fertility, love and peace. After being invisible for 14 days, Venus emerges for 245 days as the Morning Star, the "bringer of light", representing Athena, the Goddess of war and the hunt. Venus is then invisible for a further 78 days and completes a cycle of 1,460 days, exactly 4 years.

- The points of Venus' pentagram indicate 5 different groups of stars or constellations.

- The 4-year cycle of Venus was used by the Ancient Greeks to measure the Olympiads, and this is why the modern Olympic Games are held every 4 years.

- 1,460 solar years is the basis of the Egyptian Sothis' year, belonging to the God Seth-Sirius and the Goddess Sothis (or Venus).

Armed forces

Venus' 2 appearances encompass 2 divine powers of opposing forces, namely love and war. This is why the pentagram is regularly used as a symbol by the armed forces in both the West and the East and all officers in modern armies display a number of pentagonal stars on their uniforms. Such stars also appear on military machines such as planes and tanks and they even featured on some of the crusader knights' coats of arms during the Middle Ages. During the 16th, 17th and 18th centuries the pentagon was common as a design for fortresses. This is still the case today, as seen in the design of the military headquarters of the Pentagon in the USA (below left).

The Pentagon, USA

USA army tank decoration

Star fort (by Robert Fludd)

Morning Star or Eastern Star

The Morning Star also represents the Star of Bethlehem. The heart sometimes appears with the 5-pointed star, representing happiness and favourable opportunities.

Left **Various examples of the Morning Star**

Achievement

Stars are a measure of acclaim and achievement and we give stars to children for doing something well and talk of pop "stars" and sports "stars". Contrasting with its association with war, the star pentagram is a goal to aspire to.

Pentagram in magic and mysticism

The pentagram is used a symbol of faith by Wiccans and it has close associations with "magic" and mysticism. But it is also used in freemasonry and other beliefs, as in Christianity, representing the 5 wounds of Christ.

Below from left to right **Pagan pentacle; Pentacle of the Art as in** *The Sixth Book of Moses*; **the earliest-known pentagram ring (Italy c525 BC); Heinrich Cornelius Agrippa's** *Libri tres de occulta philosophia* **(15th century)**. In Renaissance occultist works, the pentagrammaton (5-letter divine name) could be arranged around a pentagram. Symbols of the Sun and Moon are central; the other 5 classical "planets" are around the edge.

Great Stellated Dodecahedron

In three dimensions of space the pentagram looks like a 20-cornered star, called the Great Stellated Dodecahedron, where each pentagonal face is capped with a pentagonal pyramid composed of 5 Golden Triangles.

Right **Robert Webb used Great Stella software to create this image** (www.software3d.com/Stella.html).

Ideologies

As a symbol of ideologies, the pentagram appears alongside other symbols to denote different creeds and belief systems. It is used in the flags of some 60 nations; for example, with the hammer and sickle it denotes Communism and with the crescent Moon, Islam.

Islamic symbol

Solomon Islands flag

Angolan flag

Algerian flag

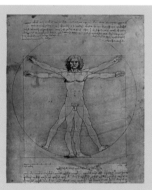

5-fold man

The 5-fold man (having 2 hands, 2 feet and a head) exists between heaven and Earth and Leonardo da Vinci's image shows hand and foot positions Squaring the Circle (see page 31).

Left *Vitruvian Man* (c1487) exemplifies the blend of art and science during the Renaissance, Leonardo da Vinci's avid interest in proportion and his attempts to relate man to Nature.

Interlaced pentagons

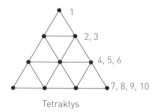

Tetraktys

Number *10*, the Decad

As described in Chapter 4 (see pages 40–41), the Triangular Tetraktys was considered by the Ancient Greeks as the perfect form, containing all that is created and *10* was the most holy number of all. The relationship between the first numbers: *1, 2, 3* and *4* was regarded as especially significant, since they totalled *10*. Number *10* is a whole that is far greater than the sum of its parts, embracing all the archetypal numbers of *0, 1, 2, 3, 4, 5, 6, 7, 8* and *9* that organize and give purpose to the Universe and life. Pythagoreans regarded *10* as a number of both fulfilment and new beginnings. In Islam number *10* is regarded as the limitless light and the Divine Essence, as it takes us beyond number itself. Likewise, many myths and belief systems use *10* with the same meanings.

As part of the decimal number system the zero is not "nothing", as it gives us the concept of many and the ability to count beyond *9*, since *0* holds the value where it is positioned as 10s, 100s, 1,000s and so on. This idea is the same as the apparently empty Circle (the same shape as zero) being a vessel for containing everything that is created. We need *0* to reach *10* and include all our fingers; this is no accident, but a clue. Time is cyclical and infinite and this is best thought of using the *0*; we cannot go beyond it because we always return to it when we count. Numbers are also cyclical, rather than linear, in their nature; a feature emphasized by the fact that their geometric forms fit within a Circle. At *10* the cycle is completed as *1 0* is the reverse of *0 1*, bringing us back to the individual Point and then the Void: *10* = 1 + 0 = 1. Multiplying a number by *10* brings that number to a higher level, but leaves its power and purpose unchanged.

- The power of *10* stems from the generative ability of *5* and the Phi Ratio. 5 + 5 = 10.

- 10 = 1 x 2 x 5 highlights the fact that the decagon (see below) can be constructed through the doubling of a pentagon in a Circle.

Above **5 and 10 are a common occurrence in fruit segments and seed pods. The passion flower has 5 petals extended by 5 brackets and it has 10-fold mathematics in its fruit and seed pods.**

Powers of 2

2 to the power of 5 = 2 x 2 x 2 x 2 x 2 = 4 x 8 = 32 and 3 + 2 = 5

2 to the power of 6 = 2 x 2 x 2 x 2 x 2 x 2 = 32 x 2 = 64, where 64 is simultaneously a Square of 8 and cube of 4, and 4 x 8 = 32

Now 6 + 4 = 10 = 5 + 5.

Two 5s represent the two complementary Spirals that work harmoniously together and are seen in the infinity symbol. This union of balanced duality that is constantly changing and merging lies at the heart of life.

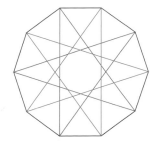

Above **Two pentagrams (red and green) within a decagon (purple).**

Creative hands

Our 2 hands of 5 fingers each, with the Duals of left and right, symbolize the Universe in its entirety and the numeric principles organizing it. To understand these numeric principles is to understand our Universe. We carry this knowledge around with us every day within our own hands; hands that are the most versatile and structurally creative part of our body, with which we in turn create (whether writing a story or painting a picture).

Top right **As we open our hands our fingers unfurl along a Spiral path.**

Right **Our 10 fingers and 2 hands form an empty vessel to be filled, just like the Void and 01...10.**

Above **An example of 10-fold symmetry in a crab.**

> 10 is the higher unity wherein the One is folded.
>
> PYTHAGOREAN DOCTRINE

Generative 10
Egyptian "sed" festivals, to renew the spiritual power of the pharaoh, occurred every 30 = 3 x 10 years. Egyptian glyphs for number 10 used the arch, also the glyph of the penis, to emphasize the generative power of 10.

"It is only necessary to make war with five things; with the maladies of the body, the ignorances of the mind, with the passions of the body, with the seditions of the city and the discords of families."

PYTHAGORAS, GREEK PHILOSOPHER AND MATHEMATICIAN (C570 BC–C495 BC)

Above **Many roses have 10-fold symmetry and the rose is a symbol associated with generation, purity and the heart. As flowers blossom by unfolding, this has caused them to be used as symbols of unfolding spirituality and consciousness, much like the lotus in Buddhism.**

The rose and Cross
Symbolism uniting the rose and the Cross characterizes the redemption of man through the union of his lower temporal nature (ego-self) with his higher eternal nature (Higher Self).

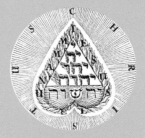

Above **The Luther Seal symbolizes Lutheranism (designed for Martin Luther in 1530).** A white rose symbolizes spirit, surrounded by a blue sky for joy. The black Cross in red symbolizes belief from the heart. A golden ring is the eternal blessedness of heaven that is beyond all joy and things, just as gold is the most valuable metal.

Above This interesting combination of symbolism is the emblem of the 19th-century Ordre Kabbalistique de la Rose-Croix society, founded in 1888. It incorporates a version of the Rosicrucian Cross with the Hebrew Tetragrammaton divine name on the arms and the Pentagrammaton in the inner pentagram.

Above **Symbol of early 17th-century mystic Jakob Böhme Christus, Iesus (Jesus) and Immanuel surround an inverted heart, for the number 5. It contains a Tetraktys made out of the Hebrew letters of the Tetragrammaton, and at the bottom, the Pentagrammaton.**

Left The red of this Rosicrucian Rose or Rosy Cross symbolizes the blood of Christ and personal sacrifice. The golden heart concealed within the Centre of the rose is the spiritual gold hidden within human nature. 10 petals is the perfect Pythagorean number.

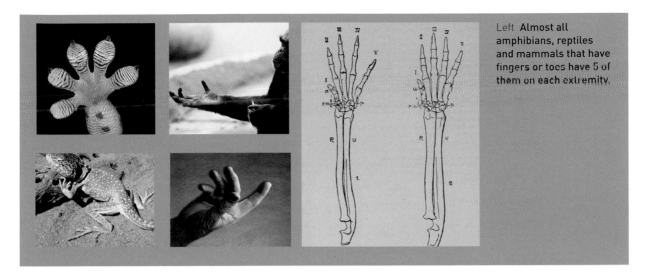

Construction and deconstruction

What would happen to the process of creation if a fourth step, or more, were taken after the third step (as described on page 40)? As part of the balance of creation and dissolution it makes sense that the steps are in reverse. This model reflects the view that the Universe and all of matter (and spirit) originated from, and will eventually return to, a single concentrated Point. The process of dissolution involves taking a further 3 steps when returning the cosmos to its primal state.

By representing these 3 steps out and to return as a model the Tetraktys (Third-level) Triangle is reversed, or inverted. It retains the Rhombus that was constructed within the primary vesica piscis, shown to the right.

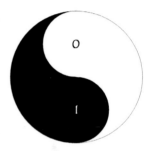

Above **The main numbers are contained in the range 01...10 and the T'ai Chi symbol can be viewed in this way when the seed dots are considered as a 1 and 0.**

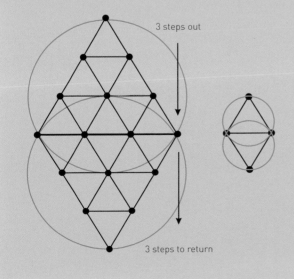

3 steps out

3 steps to return

8 Pattern-sharing and fractal geometry

I believe the geometric proportion served the Creator as an idea when He introduced the continuous generation of similar objects from similar objects.

JOHANNES KEPLER, GERMAN MATHEMATICIAN,
ASTRONOMER AND ASTROLOGER
(1571–1630)

Above **Detail from the fractal Mandelbrot Set.**

Top and above **Snowflakes, with their hexagonal structure, are a superb example of the infinite variety created by the fertile number 6. Every snowflake is unique, just as human beings are.**

THE BLUEPRINT OF REALITY COMPRISES DRY, PRECISE GEOMETRIC RULES that underpin every form of Being. Yet from perfection come evident imperfections and a vast array of physical manifestations that continue to evolve and shape themselves into a seemingly endless range of varieties and variations of form.

"Harmony makes small things grow. Lack of it makes great things decay."

GAIUS SALLUSTIUS CRISPUS (SALLUST),
ROMAN HISTORIAN AND POLITICIAN (86–35 BC)

Sharing on every level

Pattern-sharing facilitated by the dynamics of the Spiral is part of the generative process resulting in the myriad forms of Being evident on the physical plane. Even though a pattern is a predictable, recurring theme generated out of a set of parts, such as a combination of geometric shapes, what is truly incredible is how infinite variety is possible through finite rules. Looking back in time this pattern-sharing stems from simple combinations of atoms, through more and more complex stages that have built up to this point, where we are witness to the wonderful variety of physical forms evident on Earth. At every level, in every facet of reality, we can search out the patterns. Patterns feature in social behaviour, in sound, the motion of the planets, the structure of cells and even in our own creations. It is strange to think that symmetry and patterns, as in the human face, are actually approximate rather than absolute and precise mathematically. Yet the impact of the approximation has no impact on the underlying physics of the form as we see it. Only the precise symmetries and patterns of larger scales of space and time play a more important role in shaping reality.

Basic patterns from tessellations

Tessellations are based on repetition and periodicity. They are the most basic pattern-forming processes linked to the number 4 and to Squares. A single part is combined and duplicated without any modifications being made to it. The word *tessella* means "small square", from *tessera*, "square", which in its turn is from the Greek word for "four". From this comes the term "tiling", which is the application of tessellations, often made of glazed clay.

- **Regular tessellations** are highly symmetrical. Only 3 regular tessellations exist made up of equilateral Triangles (3), Squares (4) or hexagons (6).
- **Semi-regular tessellations** use a variety of regular polygons, of which there are 8. The arrangement of polygons at every vertex point is identical.
- **Edge-to-edge tessellations** are even less regular. In these, adjacent tiles should share full sides.

Regular Triangular tessellation

Regular Square tessellation

Regular hexagonal tessellation

Right **Three instances of the same tessellation patterning: in dry earth, water reflections and a giraffe's coat.**

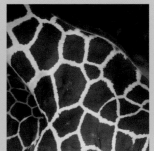

Pattern-sharing

The main types of pattern-sharing are: rotation, reflection, translation and scaling. Combinations of these are common. For example a "glide reflection" is the combination of a reflection and a translation along a line. Patterns can be seen in two and three dimensions, for example in the plane and the solid.

- **Rotations** "spin" the part around a fixed Point called the Centre of Rotation. Any rotation is possible, yet 90° and 180° are the most common. Rotations in three spatial dimensions use both a line in the plane of the Square and the Centre of a Circle, known as a "hub", enabling spin and the Spiral to form.
- **Reflections** are symmetries that transform a part into its mirror image. A two-dimensional reflection requires an axis of reflection; a line. A three-dimensional reflection requires a plane.
- **Scaling** enlarges or reduces the size of the part uniformly, or not, in all directions.
- **Sharing** in patterns is a side effect of scaling non-uniformly.
- **Translations** move the part on a plane and can be repeated infinitely.

Above and four pictures right **Examples of rotation.**

Symmetry and balance

Symmetry means "to measure together" and its opposite is asymmetry. Symmetries appear on every scale of life; in the anatomy and structures of plants and animals, growth patterns, in the way that life organizes itself and interacts socially, in patterns of movement (such as swarming bees, migrating birds and energizing dances) and even in our social interactions. In symmetrical social interactions we are all the same, as in ordinary peer groups such as classmates. Asymmetrical relationships and perceptions, such as "I am better/worse than you", are not balanced and are

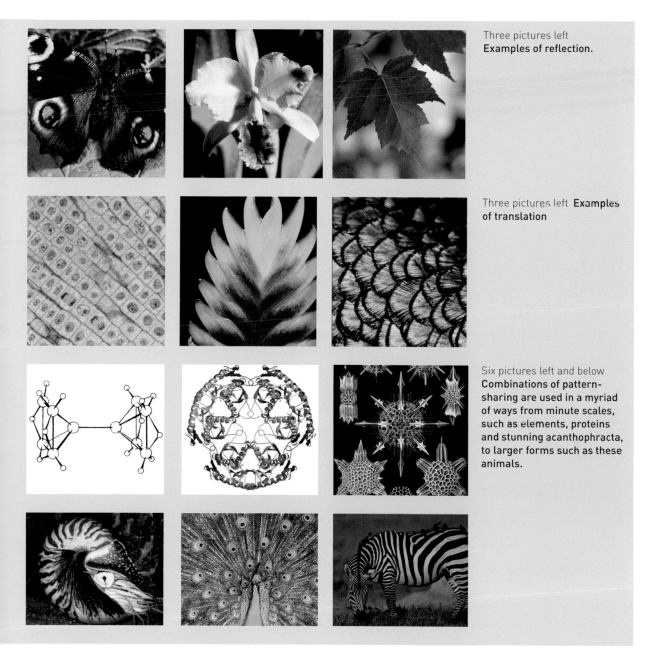

Three pictures left
Examples of reflection.

Three pictures left **Examples of translation**

Six pictures left and below
Combinations of pattern-sharing are used in a myriad of ways from minute scales, such as elements, proteins and stunning acanthophracta, to larger forms such as these animals.

power-based. Neither is better or worse; each is part of the interplay of Duals, so that choice and change can occur. We have an innate sense that symmetry brings stability, order and perfection to our experience of reality. This is why we feel stable and in control in an environment made using patterns and symmetry. Discordant surroundings are unsettling, so whenever humans create they usually strive for symmetry and order. Our affinity with symmetry and our sense of its purpose to provide order and balance in part explains why we endeavour to use perfect symmetry in symbols, architecture and most artistic endeavours.

Double Spirals in Nature

Patterns formed by 2 Spirals moving in opposite directions are prevalent in Nature, especially flowers. A typical example is the way sunflower florets grow along logarithmic, equi-angular Spirals moving in opposite directions.

Far left and left **Pineapples and dahlias are examples of double Spiral formations.**

Below left to right **Leaf, branches, overall tree and its roots.**

Scaling and pattern-sharing in a tree

Trees show us self-similarity and scaled fractals from the patterns seen in their roots and leaf veins, into the formation of branches and twigs.

Defining fractals

A fractal is "a rough or fragmented geometric shape that can be split into parts, each of which is a reduced-size copy of the whole". This property is called "self-similarity" and is derived from the Latin *fractus* meaning "broke" or "fractured". Fractal patterns also use magnification, or scaling, giving an effect known as "self-similarity" or "scale invariance".

Fractal geometry is a term originally coined by French mathematician Benoît Mandelbrot in the late 1950s and his work was published in *The Fractal Geometry of Nature*. He studied irregular natural phenomenon, such as clouds, cauliflowers, broccoli and tree branches, and realized that they all had some common features. Their patterns are repeated on descending scales, much like the principle of "as above so below" (see also page 112) and the basis of the Matrix of Space-Time-Being. For example, a small floret of broccoli looks like a whole broccoli head; a small rock is like a scaled-down version of a mountain; a twig resembles a tree and so forth. The more jagged the outline of the form the higher its fractal

Rotational symmetry

In rotational symmetry there is translation along an axis of rotation.
A part is rotated at an even, angular speed, while also moving at another speed along an axis of rotation, or translating it.

Combining these two motions creates the coiling angle. If you rotate the part quickly and translate it slowly the coiling angle is close to 0°. A slow rotation and speedy translation gives a coiling angle approaching 90°.

> To understand is to perceive patterns.
>
> SIR ISAIAH BERLIN,
> RUSSIAN-BRITISH
> SOCIAL AND POLITICAL
> THEORIST,
> PHILOSOPHER AND
> HISTORIAN
> (1909–1997)

geometry. The geometry is derived by equations that undergo iteration. Because they appear similar at all levels of magnification, fractals are often considered to be infinitely complex.

Julia Sets and the Mandelbrot Set

Julia Sets are named after French mathematician Gaston Julia, whose work was brought back to light by Mandelbrot in the 1980s. Julia Sets are fractals made of complex numbers and the ultimate Set is the Mandelbrot Set, said to be the most complex mathematical object yet generated from a few simple rules. It is fascinating to trace the fractal's contours and examine magnification of its perimeter. As the perimeter magnification is increased, ever-more complex shapes are found, such as vortices and Spirals within Spirals. And, most interestingly, tiny replicas of the whole Mandelbrot Set appear hidden within it.

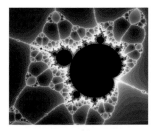

Mandelbrot Set

Below **Increasing levels of magnification in the Mandelbrot Set.**

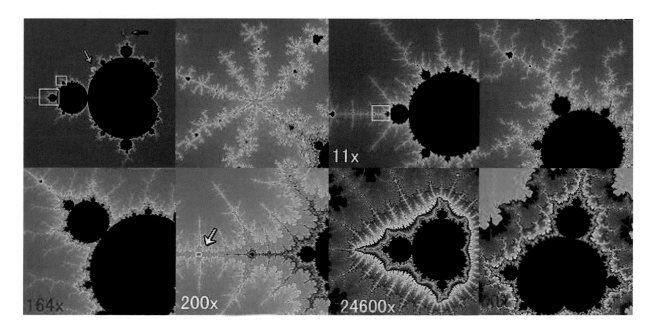

Infinite complexity

Natural objects that are generated by fractals to a degree include clouds, mountain ranges, lightning bolts, coastlines, snowflakes, various vegetables and patterns on the surfaces of plants.

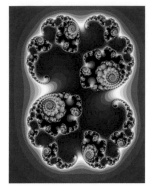

Julia Set

Koch Curve

One of the simplest fractal shapes is the snowflake curve, or Koch Curve. A line is divided into 3 equal parts and 2 sides of an equilateral Triangle are used to replace the central section. This process is repeated on smaller and smaller scales until the snowflake is formed.

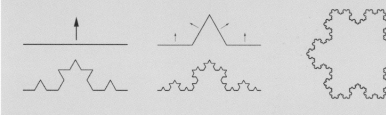

Above **Coloured Koch curve** including a Spiral

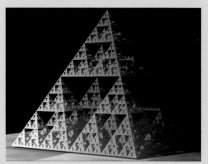

Above and right **The Sierpinski Triangle (or Arrowhead Curve) is another well-known geometric fractal pattern.**

Above **This "Mandelbulb" is a computer-generated three-dimensional fractal made using Visions of Chaos software.**

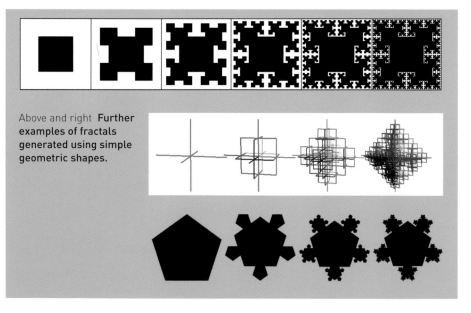

Above and right **Further examples of fractals generated using simple geometric shapes.**

Trees

A plane fractal is constructed from Squares and named after Pythagoras because each triple of touching Squares encloses a right-angled Triangle in a configuration traditionally used to depict the Pythagorean theorem.

Above **Fractal comprising Squares**

Above right **Computer-generated trees made using a varied fractal called the "dragon curve".**

Right **Dragon tree.**

The generative Spiral and infinite fractals

Combining the Spiral self-generation of patterns with the precision in growth imposed by the Phi Ratio opens up a world of phenomenal creations. As soon as the Spiral is involved, an infinite world of fractals is created, where all geometric building blocks of reality exist simultaneously, distributed in orderly style.

Scaling the Matrix of Space-Time-Being

The plant shown on the right has smaller versions contained within itself. The simple geometric patterns within the Matrix can also be nested and scaled to infinitely small and large proportions, used in numerous variations and permutations, and will still retain their essence. Every culture imitates Nature and has used the patterns of the Matrix for decorating surfaces, buildings, art and materials in decorative and sacred applications, as shown below.

Above **A cactus propagating a smaller version of itself.**

Part II

STRUCTURE OF BEING

Wisdom and Spirit of the Universe!
Thou Soul that art the Eternity of thought,
That givest to forms and images a breath
And everlasting motion, not in vain
By day or star-light thus from my first dawn
Of childhood didst thou intertwine for me
The passions that build up our human soul,
Not with the mean and vulgar works of Man,
But with high objects, with enduring things –
With life and nature – purifying thus
The elements of feeling and of thought,
And sanctifying, by such discipline,
Both pain and fear, until we recognize
A grandeur in the beatings of the heart.

WILLIAM WORDSWORTH,
ENGLISH ROMANTIC POET (1770–1850)

Part II – Structure of Being – explores how all entities, including ourselves, come into Being. What are we made of? How do we, and everything around us, exist in space and time? How do we know we exist? Are there other dimensions to reality and, if so, what role do they play? These are big questions and I want to show you how geometry, science, art and the insights of spirituality can help you on your path to a greater understanding of reality.

9 To be and Being

Great indeed is the sublimity of the Creative, to which all beings owe their beginning and which permeates all heaven.

LAO-TZU, CHINESE PHILOSOPHER (6TH CENTURY BC.)

THE CIRCLE, VESSEL OF THE ABSOLUTE VOID, is undifferentiated, time-less and space-less. Pervading the Void is the potential "to be" anything imaginable. With the process of creation and pattern-sharing, made possible through the blueprint of number and geometry, vibrations undulate through the Void with a symphony of different frequencies to give birth to different forms of Being; Beings of varying degrees of matter and spirit, such as humans, rocks, oceans, stars, gases and Beings of finer vibration that exist more in spirit. The spiritually inclined believe that all forms of Being exist in the realms of heaven and Earth at a plane according to their vibration. The "higher" the Plane of Being, the finer and more subtle the vibrations are. Lower down the Planes of Being vibrations are much slower and the form of Being more "solid". Consistent with the principles of duality no Beings are ever static, but continually changing from one condition into another.

Connection of the Planes of Being

Ignoring the possibility of the existence of the Planes of Being limits the understanding and full appreciation of the true essence of Being. Consider the human body. We understand the physical mechanics well, but many people can see

Above **Abstract conceptualization of Being crystallizing out of non-Being.**

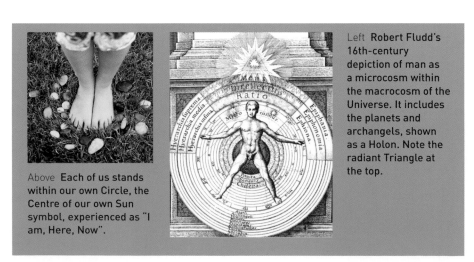

Above **Each of us stands within our own Circle, the Centre of our own Sun symbol, experienced as "I am, Here, Now".**

Left **Robert Fludd's 16th-century depiction of man as a microcosm within the macrocosm of the Universe. It includes the planets and archangels, shown as a Holon. Note the radiant Triangle at the top.**

Jacob's Dream

According to the Old Testament of the Bible, Jacob dreamed that there was a ladder on Earth, the top of which reached to heaven and God and that angels ascended and descended. According to Christianity this means that a continual intercourse between heaven and Earth should take place. Islamic interpretation of this story is that the ladder is the path of creation back to its beginning. This story also closely relates to teachings based on the Tree of Life, such as the Kabbalah.

I cannot express it: but surely you and everybody have a notion that there is, or should be, an existence of yours beyond you.

EMILY BRONTË, ENGLISH NOVELIST (1818–1848)

colourful auras surrounding the forms of living Beings (human, animals and plants). A few can even look inside to view internal organs. So there is more to Being than most people can physically see.

Many of the latest theories in physics are beginning to conclude that there are Planes of Reality, where everything is connected beyond the tangible, physical form. This is conveyed using terms such as "events are connected", or "correlated" in these higher Planes. Quantum physicists use models that are almost identical to spiritual ones to explain the nature of matter beyond the atomic level.

Above **Nepalese Buddhist Triangular tower atop a dome on a temple. Note the eyes looking out.**

Hierarchies and the unified whole

In some belief systems, such as Christianity, the Planes of Being are regarded as separate worlds, with a specific order and ranking. Suitably, Triangles are used to organize many hierarchies of life, for example the food pyramid, the animal and plant kingdoms and Maslow's Hierarchy of Needs. Indeed life appears to favour multi-levelled structures of systems within systems.

In geometric Triangular hierarchies *1* is placed at the apex. Below *1* the Triangle unfolds and makes larger Triangles, just as occurs in the Triangles of *3*, *6*, and *9* that are at the basis of the Grid of Being (see page 41). The lower levels represent increasing differentiation, as can be seen in the evolution of species. The higher levels depend upon the broad foundations of the lower levels for their survival. Without these "lower" foundations of Being, the whole Triangular structure crumbles back to unity. Unfortunately, this type of model is taken to imply superiority and inferiority of forms of Being and Planes of Being relative to one another because of the interpretation of "hierarchy". More appropriately, multiple levels of any such model should be viewed as parts of a totally unified whole; as highlighted by the symbol of a Triangle within a Circle,

Above **A 12th-century vision of the angelic hierarchy as radial Circles.**

Above **Spires connect us with the heavens – very much like radio antennae. Their pinnacle is also effectively set in the Centre of the Circle and the source.**

Things derive their being and nature by mutual dependence and are nothing in themselves.

NAGARJUNA,
ANCIENT BUDDHIST SAGE

Cappadocia, Turkey

Triangular mountains

Regard this common feature of a Triangular mountain as a peak supported by broad foundations, set as a pinnacle within a Circle. This is a natural phenomenon that the humbler, yet stunning, rock formation in Turkey (see left) mimics on a smaller scale.

Rocks break off the mountain and tumble to the ground, surrounding it, gathering momentum and setting off other incidents on the way down. On higher levels there is little life and as you move downward greater variety of life appears on the slopes. Eventually exposed pinnacles are eroded to merge with the flat surface of the spherical Earth.

Life cycles

Hierarchies can be represented geometrically in the form of Triangles within Circles. A Circle for the eternal life cycle and a Triangle for birth, life and death resulting in new life. Leaders may be viewed as being at the top of a hierarchy or at the Centre of a Circle. As such the impact of their Intent radiates out from their "inner" Circles to those on the rim.

Far left **Japanese Popillia beetle life cycle.**

Centre **Giant woodwasp life cycle: larvae, pupa and adults.**

Left **Giant galactic nebula NGC 3603. The various stages of the life cycle of stars depicted in a single image.**

Concentration and transmutability of energy

Vibrations in Being concentrate as they approach the pinnacle of a Triangle. At the top is the Point, *1* and stillness. Viewed from the top we see this as a Spiral, with tighter coils as it closes in on the elusive still Centre, like the "eye" of a storm. Take the food chain as another example. Lower levels of vegetation are plentiful and diverse, whereas the food sources at the top are limited, but highly concentrated. The energy supplied by food has literally moved from the bottom to the top through the food chain, in the process becoming more complex and changing its form from a vegetable into an animal.

Such examples in Nature reinforce the principles that vibrations within energy move from the base to the apex in Triangles and that energy concentrates itself in the process. The food pyramid model also demonstrates the transmutability of energy, in that everything has the ability to transform, to change its state up and down the Triangle. Note that energy moving from the base to the apex concentrates itself in the process.

Above **The Triple Moon, symbol of the Triple Goddess. This is a symbol of the 3 phases of the Moon, the 3 phases of human life and the cycle of all things. As a symbol of the cosmic cycles it is also a representation of the process of life and death, reincarnation and rebirth.**

Evolution of the whole

As part of an interpenetrating grid of relationships, each Plane of Being needs to be considered in the context of the mutual relationships that sustain the whole. Networks within networks sustain Being on every level in a flux of perpetual motion. We can see how evolution relies on a large body of shared learning, so to reach the pinnacle would be a jointly acquired culmination in the evolution of the whole. The pinnacle can only be reached by moving up the layers supporting

From wonder into wonder existence opens.

LAO-TZU, CHINESE PHILOSOPHER (6TH CENTURY BC)

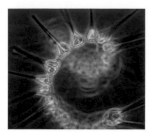

Above **Phytoplankton are the foundation of the oceanic food chain, supporting a myriad of life forms.**

Above **A pupa, wherein a caterpillar transforms into a butterfly.**

Above **Ladders, even corporate ones, rely on the strength provided by a Triangle. Reaching the pinnacle by moving up the steps is regarded as an achievement based on hard work, skill and knowledge, but it is easy to fall back down to the edge of the Circle.**

it, since each step becomes the foundation for the next step. In a spiritual journey these principles are upheld as a process of growth where the ninth step is the culmination before perfection of number *1*.

Linking the Square and the Triangle

The growth and evolution of each form of Being influences the Circles they move in and ultimately the growth and the evolution of the Universe as a whole. In the human experience we are Dual Beings (of matter and spirit) "born" to transform our state of Being, evolving in body and soul as a result of our experiences on "four-square Earth". All life experience and progress is facilitated by the Spiral Life Force, moving evolution of the Whole upward and inward to the Centre, to be returned to a condition of Unity. And so the Circle, Square and Triangle serve to facilitate this through their structure, reminding us of this as symbols and serving as tools for us to work with space, time and Being.

The Allegory of Age Governed by Prudence by Titian (c1565–1570). A Trinity depicts the 3 main stages of manhood: youth, maturity and old age. This painting is thought to show the artist, his son Orazio and a young cousin Marco Vecellio.

The Life and Age of Woman – Stages of Woman's Life from the Cradle to the Grave (c1849) illustrating 11 chronological stages of virtuous womanhood (with the 30s considered to be her peak years). Note the ascent and descent, or waxing and waning, in her life cycle, from birth into maturation and then decline before death.

10 Mind and sentience

Cogito ergo sum.
I think; therefore I am.

RENÉ DESCARTES, FRENCH PHILOSOPHER
AND WRITER (1596–1650)

Cogito cogito ergo cogito sum
I think that I think; therefore I think that I am.

AMBROSE BIERCE, AMERICAN AUTHOR
AND SATIRIST (1842–1913)

MIND, BODY AND SPIRIT; THESE 3 PARTS OF THE HUMAN CONDITION are a key Trinity. Like the image in a mirror, our ethereal and intangible, spiritual part exists outside the realm of any physical tools of measurement, while the physical human body we describe in general terms as measurable matter.

Humans are essentially Dual Beings, while the third condition of the Mind sits in the middle, straddling the physical and spiritual aspects to unite and balance them. The Triangle and the numbers 3, 6 and 9 keep recurring in science, Nature, art and beliefs as models, symbols and tools to help us understand and work with everything to do with understanding Being and what it is to "become". And it is in areas as contentious as the nature of Mind and consciousness that they are particularly useful.

Above **Disciplines such as yoga teach understanding and control of the Mind, body and spirit.**

The Mind

The Mind is essential to the perception that we exist and to the formulation of the statement "I am" and any definition of what it means "to be"…to exist. Our Mind and its processes have both measurable and immeasurable components and, sadly, current models tend to be limited by physical parameters and descriptions trying to measure the form and scope of awareness.

There are mathematical models of consciousness that cannot be argued in finite terms and proponents of these models conclude that consciousness is infinite. This view is supported by mystics, who use the number 9, the number of infinity, to symbolize the unending expansive nature of consciousness. Furthermore, Hindus, Buddhists and other mystics maintain that the infinite ocean of energy filling the Absolute Void comprises pure consciousness. Since all Beings of matter and spirit manifest out of this ocean, logic says, again, that all Beings must have a degree of consciousness.

> "I" translates
> as AHAM
> A = Siva,
> HA = Sakti,
> M = union
>
> HINDU SEED MANTRA

Above **Conceptualizing Mind as only a brain "boxes" us into a linear world of individuals rather than as being interconnected to all Beings.**

Brain and Mind

One strand of scientific thought argues that consciousness is transcendent and beyond understanding. Another strand places consciousness in the physical world of brain activity. This school of thought argues that only "living" animals have consciousness and this is quite contrary to the belief that absolutely everything originates from pure consciousness. At this end of the spectrum, the emphasis of research on consciousness is placed on the neural activity associated with the concept of "I" in the cerebral cortex, located at the front of the brain behind the forehead. This is the area most associated with the generation of conscious awareness.

Indeed brains are very sophisticated, but do they equate to the Mind? Our capacity for thought, the power of intent and the boundless realm of our imagination hint at a much wider definition of Mind. Plants may have a Mind, but no brain, so Mind and matter are not separate but different aspects of the same phenomenon of life. Following the same lines of reasoning, as in mysticism, if all things originate in the original pool of pure consciousness, then it is not realistic to equate consciousness with the brain alone. A brain is a tool for our Mind, similar to a vehicle and our Mind is not physically limited by it because it bridges spirit and matter. Mind knows our true identity, which we sense in our Point in

Folding

Folding maximizes surface area, which can be seen in numerous life forms, such as reef-forming coral, fungi and human brain grey matter. It optimizes the space available in our heads and our brain capacity.

Brain coral

Brain coral

Brain fungus

Orange puffball sponge

Puffball fungi

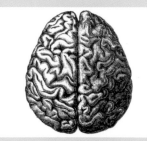

The human brain

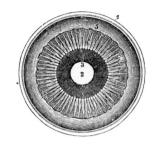

Space-Time-Being. It is verbalized as "I am". Our "I am" is listener, watcher and witness to the foibles of our physical ego-self that we identify with as in a physical body. By identifying with ego it is set into action, makes incessant demands and veils our Mind.

Perceiving Being, I and Eye

On which Planes of Being does a form become aware of itself, of other forms of Being, perceiving a reality and how is this sentience to be measured? Once again we look to Triangles to help us define structure and define the aspect of aware-ness of existence experienced as "I am". In English "I" means "me, the subject or object of self-awareness" or the recognition of one's existence. Significantly, in English "I" is pronounced in the same way as the word "eye". Sentience is symbolized by the Eye. The Eye is also a measure of conscious awareness and the scope of a Being's ability to see reality and absorb knowledge, embodied by light entering into the Eye for processing. The Eye opens, perceives itself to be in a particular location in Space-Time and existing as an entity in the Grid of Being.

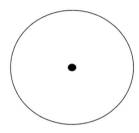

Top **Anatomical drawing of an eye, resembling many Sun and wheel symbols.**

Above **A simple Circle with a Central dot, representing the eyeball of the Universe.**

The original single Eye

Point of origin and return, the Central Point is expressed as the personal expe-rience, "I am, Here, Now". As such it may be represented by a single, still and dimensionless Point around which the cycling dynamics of the Circle operate, binding events together. It represents the ultimate form of non-manifest existence of Being that cannot be made any smaller. This is pure "I AM". Every culture uses the symbol of a Circle with a central dot. Often it is associated with the Sun or gold, but it most often represents the Centre, from which everything originates. Your own eyeball, a white sphere with a black dot in the Centre, looks like the Circle with the Point, *1* or I, in the Centre. Your eyeball is a passive receiving Eye, accepting the light to form images of your external reality.

Lunar and solar Eyes

If we imagine a single eye actively looking out it will "see" a passive reflection of itself on the horizon – in effect, its Dual. These Dual Eyes remain necessary parts of the same whole, working together, much as our left and right physical eyes do. The symbolic solar Eye radiates light, while the lunar Eye passively reflects light, just as the Sun and Moon do. Each also represents the functions of our own 2 physical Dual Eyes. These link the hemisphere of the brain (also shaped as a vesica piscis); one side for the male logic and the other female creativity. Each side contributes their view of your external reality to give you a blended picture.

"Our treasure lies in the beehive of our knowledge. We are perpetually on the way thither, being by nature winged insects and honey gatherers of the mind."

FRIEDRICH NIETZSCHE, GERMAN PHILOSOPHER, POET, COMPOSER
AND CLASSICAL PHILOLOGIST (1844–1900)

> Simply the thing I am shall make me live.
> WILLIAM SHAKESPEARE, ENGLISH PLAYWRIGHT (1564–1616)

Above **Early 18th-century Eye of Providence (Masonic symbol).**

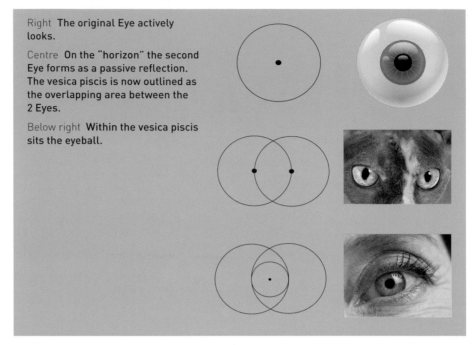

Right **The original Eye actively looks.**

Centre **On the "horizon" the second Eye forms as a passive reflection. The vesica piscis is now outlined as the overlapping area between the 2 Eyes.**

Below right **Within the vesica piscis sits the eyeball.**

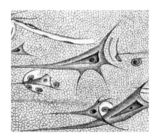

Above **Triangular nerve cells in the human brain.**

Above **Consciousness "frozen" in physical form is often represented as an Eye within a Triangle.**

Eyes of Horus
The right, white Eye of Horus is the Sun (Ra) Eye of the sky god. It is associated with the left side of the brain and represents the masculine. The left, black Eye of Horus is the Lunar Eye (*Wadjet* or *Udjat*), meaning "made whole or restored". It is the feminine and right side of the brain, linked into the cycles of life and the ability to "see" time and change.

Right **Lunar and Solar Eyes of Horus on an Ancient Egyptian stone tablet.**

Seeing with 3 Eyes
In the Chinese text *The Secret of the Golden Flower*, it is written that between the Sun and the Moon (the 2 Eyes), there is a field of 1 inch square (2.5 sq cm) that is the heavenly heart, the dwelling of the light, the golden flower, where thoughts are collected together and light circulated throughout the spiritual body. Other cultures also describe the legendary capacity of the Third Eye, the Eye of insight that straddles the 2 physical Eyes and hence directly correlates with the Mind, while the 2 physical Eyes look outward for the ego-self and decide how we see ourselves physically and emotionally in comparison to all other Beings. At our

core is just Being, expressed as "I am". Depending upon how we view ourselves within reality we add multiple layers of ego identity. These layers hide our core essence and are used as reasons, even excuses, for our behaviour. For example, I AM human, British, female, middle-aged, daughter, wife, mother, author, artist, in good health and so on.

The French philosopher René Descartes maintained that the Mind exists separately from the material, physical world. His theory was that the brain is a type of receiver linking into the Mind via the pineal gland, located in the prefrontal cortex, lying mainly behind the forehead. In mysticism the pineal is the place of the

Above **2 eyes are a trait of sentient, aware animals. Could it be that they share our awareness of the solar male logic and lunar female wisdom?**

soul and the mysterious Third Eye. So our Third Eye is also an organ of sensation, but of higher perception, since it has access to information about the Universe that is ethereal in Nature. And, like our other 2 physical eyes, the Third Eye can also be opened and closed. When open and in use, the Third Eye brings forth the insightful intuition of our Mind.

Compound eyes

Even compound eyes are structured with the number 3 as their basis in their hexagonal array. This array is made of sensors, each of which has its own lens and photosensitive cell(s). Some compound eyes have up to 28,000 sensors and can give a full 360° field of vision.

Right **Two photographs of the surface of a fly's eye.**

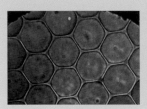
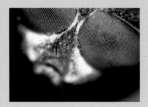

All-Seeing Eye

The imagery of an All-Seeing Eye can be traced back to Egyptian mythology and the Eye of Horus. It also appears in Buddhism, where Buddha is referred to as the "Eye of the World" throughout Buddhist scriptures and is represented as a Trinity in the shape of a Triangle known as the *Tiratna*, or Triple Gem.

This same ideal is also used in freemasonry and in the Christian faith, as an Eye within a Triangle, to convey the all-seeing nature of the Holy Trinity. Note that the Eye at the top of a pyramid with a Square base is the Centre of the cube (see page 85). This fixes our Eye Here on Earth. By viewing Earth from the top we see our life in context and can view more of the Square.

Below **The Second Great Seal of the USA** *Annuit Cœptis*, **above the All-Seeing Eye looking down with favour on the unfinished pyramid. This means "God has favoured the work."**

Below centre **The Eye of Cao Dai, the symbol of Caodaism.**

Below right **Freemasonry symbol of the Eye of Providence surrounded by sunbursts.**

Locating the Centre

The original Eye is at the Centre of the Matrix of Space-Time-Being in two and three dimensions.

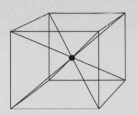

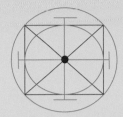

Far left A three-dimensional cube's Centre is the "top" of 6 Square-based pyramids, whose bottoms are made of each side.

Left The Centre of the cube is the common Centre of the Square and Circle with the diagonals highlighted.

- The Square's corners touch the rim of the Circle surrounding it and establish its strong foundations and a diagonal Cross.
- Another Circle within the Square makes contact with the sides where the north, south, east and west "doors" are. These "doors" are connected by the horizontal arms of the green Cross.

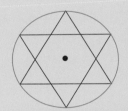

Left The Eye at the Centre of the Triangle and three-dimensional Star of David. The scope of seeing is the boundary of the Triangle and also represents how we see our ego-self in reality. Yet this is only part of our whole Self, frozen in a physical form.

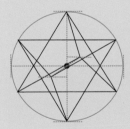

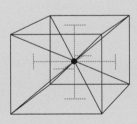

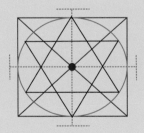

Left The Centre of the Gateway to the Heavens is "I am, Here, Now". It is the Mind that thinks, knows and experiences this fundamental and powerful thought; a thought that then becomes a reality.

Above A Cross reminds and shows us where the Centre of all the key shapes is located.

Left In this sacred geometric model (Jain Dharma) 2 Eyes at the top are part of the outer circuit of the Holon. Jain philosophy and practice advocates the necessity of self-effort to move the soul toward divine consciousness and liberation.

Above **Indian carving with the bindi (or "tilak" when worn by men) on the forehead between the eyes, which is said to be the seat of "concealed wisdom".**

Prefrontal cortex and the imagination

Our prefrontal cortex is the most "evolved" part of our brain, associated with self-awareness or "I am", self-will, making choices and finding meaning in life. Significantly it is this part of the brain that evolved dramatically in the period coinciding with the arrival of our imaginations. Here ideas are created, plans constructed and thoughts connected up with their associations, to form new memories. Our memories are then brought together into our consciousness. Emotions from physical survival systems come into the sphere of subjective feelings. By combining this information we can interpret data in a way that looks "behind the scenes", "sees" beyond the physical and "sees" hidden links.

"Our observation of nature must be diligent, our reflection profound, and our experiments exact. We rarely see these three means combined; and for this reason, creative geniuses are not common."

DENIS DIDEROT, FRENCH AUTHOR AND PHILOSOPHER (1713–1784)

Opening your Eyes

Returning to the beginning of this book (see page 10), it is our ability to imagine possibilities that enables us to look beyond the surface of our surroundings and visualize hidden things. It allows us to look at our world from creative, logical and intuitive points of view: all are metaphors based around the idea of seeing.

Without the realm of the imagination we would not be able to pose questions such as why?, when?, where? and how? Without the intellect and logic we could not reason sequentially. Without intuition we could not read behind the scenes and receive flashes of inspiration that have no logical foundation. Without these

The eye is the first circle, the horizon from which it forms is the second; and throughout nature this primary figure is repeated without end.

RALPH WALDO EMERSON, AMERICAN LECTURER, ESSAYIST AND POET (1803–1882)

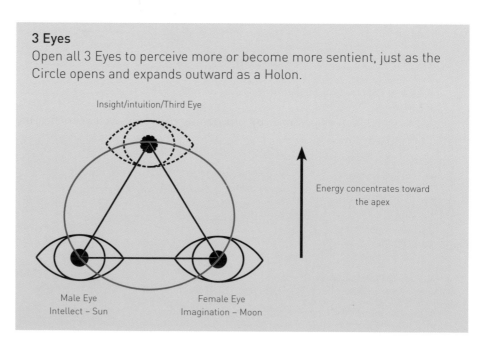

3 Eyes
Open all 3 Eyes to perceive more or become more sentient, just as the Circle opens and expands outward as a Holon.

Insight/intuition/Third Eye

Energy concentrates toward the apex

Male Eye
Intellect – Sun

Female Eye
Imagination – Moon

> **I AM**
> Hindu tradition states that Brahman says the first sound "aham",
> "I AM", made of the first, middle and end letters of the Sanskrit
> alphabet, is like three points on the Circle, forming a Triangle.

perspectives we would not be able to form an answer or communicate and debate the possibilities with others. And it is by opening all 3 Eyes to look at things from different points of view and merging them as one that we can expand our awareness of reality and our place within the Universe.

> *"The real power, the fundamental greatness....is not the physical world, not the magnitudes of Space, but the magnitudes of Time. Time, that mysterious condition of universal mind, which alone makes the ordering of the Universe in Space possible, although its own particular relations to matter are necessarily determined by material events and movements."*
>
> FROM A COMMENTARY ON THE *BRIHADARANYAKA UPANISHAD*,
> PRE-BUDDHIST ERA

Above **This ancient bronze Buddha has the physical eyes closed to focus attention and energy on the Third Eye in the Centre of the forehead.**

The human interface

Our bodies are gifts that allow us to explore reality via the human form. Our brain works our body and it is a tool for our Mind. Our senses are receptors that feed our brain and our sight is the most vital of senses. Our Eyes are our windows onto reality that allow us to see ourselves in relation to everything else manifest out of consciousness in Space-Time. And by looking out of our physical, material bodies we not only observe and explore reality, we also create our reality, since we cannot divorce the observer from the observed.

By expanding our sentience the experience of life becomes more rewarding and we become more alive and more aware. To be "alive", therefore, means using all of our senses to their full capacity, by opening every Eye to expand our awareness of reality as much as possible. By broadening the scope we experience life more fully and learn more from it.

The scale of sentience and the size of the Eye apply to all forms of Being; not just humans. They apply as much to planets, trees, space, gas... every physical form that is manifest. "As above, so below" (see pages 115), it also applies to the spiritual Planes and our ability to perceive reality beyond the physical, beyond the limitations of our physical bodies by using the Mind.

> *"Man's mind, once stretched by a new idea, never regains its original dimensions."*
>
> OLIVER WENDELL HOLMES,
> AMERICAN PHYSICIAN (1809–1894)

11 Classical Elements as metaphors

> We cannot conceive of matter being formed of nothing, since things require a seed to start from. … Therefore there is not anything which returns to nothing, but all things return dissolved into their elements.
>
> LUCRETIUS, ROMAN POET AND PHILOSOPHER (C99–C55 BC)

FIRE, EARTH, AIR, WATER AND QUINTESSENCE/ETHER are known as the 5 Classical Elements. They are dynamic metaphors for the transmutability of Being through different phases of manifestation in the Planes of Being within the Triangle.

While fire, water, air and earth are the indestructible Elements, Aristotle added ether as the quintessence: "essence of that which is unchangeable and eternal and the key to immortality". Quintessence infuses the other 4 Elements. He asserted that fire, earth, air and water were Earthly and corruptible. Since no changes had been observed in the heavens, he believed that stars must be made of an unchangeable, heavenly substance, which he called "ether". Metaphoric and symbolic ether can be likened to the Life Force, which imbues all Being.

Variations in the categories of Elements	
Greek	Air, water, fire, earth and ether
Hinduism (Tattva) and Buddhism (Mahabhuta)	Air/wind, water, fire, earth and space
Chinese (Wu Xing)	Wood, water, metal, fire, earth
Japanese (Godai)	Air/wind, water, fire, earth and sky/heaven
Medieval alchemy	Air, water, fire, earth; also sulphur, mercury, salt
Babylonian	Wind, sea, sky, earth

Mixing it up

For "life" as we know it to have emerged from the primordial soup certain ingredients were required – water (water), warmth (fire), oxygen (air) and nutrients (earth). The fifth intangible, unquantifiable and more mystical ingredient is the Life Force. These ingredients of "life" correlate to the Classical Elements and, as such, all of Nature's creations have all of the Elements within them. The magic is in the mixture. As in cooking, where we mix ingredients together in varying combinations at different temperatures to make new forms, the Elements are combined in all forms of Being to varying extents.

Above **Some of the many forms on Earth of fire, water, earth, air and ether. Our surroundings are rich in variety of scales, hue, texture and design.**

Love and hate
The Greek philosopher Empedocles proposed that all the Classical Elements existed together in fixed quantities from the beginning and were mixed and unmixed by Love and Hate. This allowed him to agree with Parmenides that Being never really changes.

Far right Fire "cooks" the ingredients of life so that they can transform their state and density. This picture beautifully depicts the process as a round pan placed within a Triangular fire.

Right In the Circular vessel the Elements combine in varying combinations to create a wide variety of Being; much like the range of breads that it is possible to make from the same essential ingredients.

Above The "alchemical" process in the uses of wheat.

Transformation of matter and Triangles

Just as the food chain demonstrates the transmutability of energy, in that everything has the ability to change its state, this same dynamic of energy can be seen in the transmutability of Being, as shown by the Classical Elements. As all Being comprises simply energy, the different Elements represent different states of vibration. So although the form of Being changes its state, its essence remains unchanged. The Classical Elements provide invaluable metaphors that allow us to describe and appreciate the attributes of these states of Being. This is why, throughout cultures, the Classical Elements are combined with universal geometric models, such as Medicine Wheels, to add another layer of understanding to the dynamics of reality and the process of creation.

"What canst thou see elsewhere which thou canst see here? Behold the heaven and earth and all the elements; for of these all things are created."

THOMAS À KEMPIS, GERMAN MONK AND WRITER (1380–1471)

The Elements in the Tetraktys

Significantly the Pythagoreans added the Elements to the Tetraktys (see page 40), which they revered. The 4 physical Elements are placed in order of increasing density, with fire at the top of the Triangle. Earth is at the bottom and is the 4-Square foundation upon which the other Elements are placed. By being presented as the Tetraktys, all continually cycling states of matter are contained with the numbers 0, 1 to 10. 9 numbers surround the Central 1, which represents the source from which everything originates and the 5th Element, which is eternal and unchangeable.

Above A change in state can sometimes be dramatic, as depicted by this illustration of Berthold Schwarz ("Berthold the Black"), legendary German alchemist of the late 14th century, credited with the invention of gunpowder.

The Tetraktys can also be placed within a Circle, thus reminding us of the eternal Circular vessel containing all forms and states of Being. As has been shown (see pages 46–61), 5 and the Spiral is the dynamic that facilitates change and transformation in expansion and contraction, growth and decay and hence creation and destruction. Decay and destruction are the Duals of creation and growth and just as essential to the process of transformation in Being.

Western alchemists

Alchemists based their "chemistry" on the understanding of the Classical Elements. Western alchemy focused on chemical transformation, which is why alchemy is still mostly associated with turning base metal into gold. Although their goal was also to discover the Philosopher's Stone, or the Elixir of Life, they believed in the transmutation of man into a form with eternal life (as symbolized by the base metal becoming gold). In a way, they were looking for spiritual perfection through the control of physical matter and its spiritual aspect.

Elements in the Tetraktys

The Pythagoreans placed the Elements at each level of the Tetraktys in order of their density.

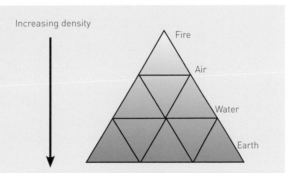

The Elements in the Holon

By representing the Elements as a Holon we are essentially viewing the Tetraktys as a cone from an aerial perspective. The further away you are from the fire the cooler and denser the form of matter becomes. Connection from Circle to Circle is provided by the Spiral Life Force, facilitating seamless motion and change from one condition to the next.

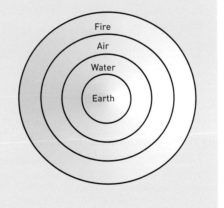

4 around 1

The traditional Elements work together in a subtle, creative dance and their union facilitates the variety of physical life that is animated by the Life Force. All Elements are unities, in that none is superior to any other and they balance each other (see page 54).

Number *4* is the Square, the 4-Square Earth and the basis of the Grid of Space facilitating the tensions of double duality: 4 = 2 + 2. The spiritual Elements, fire and air, are Duals to the material Elements earth and water, representing the union of opposite principles. For example, water (female) and fire (male) are opposites that destroy each other. Symbolically, when they combine as the descending water and the ascending fire, air (masculine) and earth (female) are created. According to myth we were originally created from clay; a mixture of earth and water. Similarly, Native Americans believe the energy of physical activity is linked to fire and air (the Sun), receptive intuitive subconscious activity to water and earth (the Moon).

The symbolic dragon

Combining the characteristics of the 4 Elements the dragon symbolizes the Sun and the Moon, masculine and feminine and the unity underlying them. Visually a dragon combines the Elements, having the wings of a bird and scales of a fish or a snake; it breathes fire and often protects a hoard of golden treasure.

Dragons are representative of the spirit of water that breathes fire, where fire has the ability to transform matter into the ethereal, as symbolized by water turning into vapour through heating.

Top **Heraldic crest.**

Above **17th-century symbol of volatile transmutation from *Theatrum Chemicum Britannicum* (1652) by Elias Ashmole (English astrologer and student of alchemy).**

Above **A Chinese example of Dual dragons within the vesica piscis – the fertile area where opposites blend.**

Above **Late 17th-century image of the "chymical marriage". This represents the union of opposites as a royal marriage between a king and a goddess; body and spirit. The product of the union is the hermetic androgyne (dragon).**

Above **Engraving from *Tripus Aureus* (1618) by Michael Maier, a German alchemist, from a summary of the preparation of *The Philosopher's Stone*: 3 crowned serpents symbolize the fundamental substances sulphur, mercury and salt. The dragon is at the Centre of a Triangle, with wings growing, within a vesica piscis.**

> **The Cross**
> For most Ancient peoples and alchemists the arms of the Cross represented the 4 indestructible Elements of air, water, fire and earth. Notably, the Central Point symbolized the fifth Element ether, *quinta essentia*, the Life Force. In the Middle Ages alchemists appended a Cross to signs for chemical elements and compounds. The Cross is also associated with the Cardinal Directions and this has significance in many geometric models when the Elements are linked to the Directions.

Above **The Ouroboros (the symbol of a dragon eating its own tail) was also used to represent transformation of matter. From** *Atalanta Fugiens* **(1617) by Michael Maier.**

Ritual and symbolism

Western alchemy originated from the Egyptians and Greeks, who focused on physical and practical manifestations. Empirical and practical at first, by the 4th century AD astrology, magic and ritual had begun to feature in their work and hence the power of intent. Every alchemical act was recognized as being under the influence of universal forces (such as atmospheric and astrological alignments) and because of this alchemy was conducted according to strict timing.

Extensive use of symbolism embodies the mystical aspect of alchemy. For example, the Circle represented the eternal cycle of Nature and the symbol of transforming matter is the Ouroboros, a dragon eating its own tail (see illustration right). The central axiom of alchemy was that "the vessel is one" (*Unum est vas*). The vessel used during alchemical experiments must be completely round to imitate the spherical cosmos of the Void, so that the influence of the stars may contribute to the success of the alchemical operation.

During the 17th century the split between the physical and mystical/spiritual aspects of alchemy started. Physics, chemistry, medicine and astronomy became the physical sciences and all other aspects were pushed into the background. Even so, we find that the great scientists of our age, such as Sir Isaac Newton, Robert Boyle and Albert Einstein, were attracted to esoteric aspects. Mysticism is vital if reality and the true nature of Being are to be studied and comprehended.

The 5 Chinese Elements

The Eastern Elements, such as the Chinese Wu Xing, are regarded as more figurative, but they are still important as they highlight the integration of the individual within Nature and the Whole. In Chinese alchemy the main focus is on the panacea and Elixir of Life. Chinese alchemy is not about "having" but about "being". It is linked to the Chinese philosophy based on happiness, being bound to finding your Inner Self and not in others. Early Taoists searched for the "isles of the immortals"; herbs or chemical compounds that could ensure immortality. Their endeavours advanced herbal medicines, pharmacology, a healthy diet, massage and movement for keeping the body youthful.

The Chinese Elements are linked with Duals of Yin and Yang. When Yin and Yang interplay they give rise to fire, earth, water, wood and metal. These

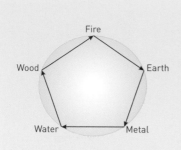

The 5 Chinese Elements

Spring – Wood Element
Wood bursts forth as new, vigorous growth and expansion. Spring is the time of organization and seeding ideas.

Summer – Fire Element
Fire reflects the brilliance and power of the Sun at its highest. Summer is the season of maturation.

Late summer – Earth Element
Late summer can be humid and damp, like the earth. Our summer efforts are "harvested" and we can enjoy the "fruits" of our labour.

Autumn – Metal Element
Autumn is the season of change and change is a word deeply embedded in Taoism. When change is not present there is stagnation and death. Although leaves wither, seeds of the future are distributed.

Winter – Water Element
During winter months the days are short and nights are long, which compels us to slow down. It is not time to deplete reserves, but rather a period of storing energy.

Above **The Pagan pentacle also represents the 5 Elements. The apex at the top of the star within the Circle represents the spirit of God that manifests itself in the emanation of the many deities and in the nature of the Universe. Its use in "magic" is not unlike that in alchemy, based around the use of both geometric symbols and rituals.**

Elements are used to describe the abstract forces that exist within and beyond the physical. The nature of the Chinese Elements is clearly illustrated in the eternal Circular cycle of the seasons. By living in harmony with the seasons (climatic changes, the mood, the colours, seasonal and regional foods) we attune our Self to the great Tao, which is the natural order and path of the Universe, and feel more at one with our environment. Chinese acupuncture extends this connection into personality type and our internal organs. The Chinese Elements' perspective of Being considers the condition of the individual Being within the whole.

Tibetan philosophy and Dzogchen

As in the Chinese view of the Elements, in Ancient Tibetan philosophy, or *Bön*, the 5 Elemental "processes" of earth, water, fire, air and space are the materials of all phenomena that physically exist (also known as "aggregates"). They form the basis of the calendar, astrology, medicine and psychology and are the foundation of the spiritual traditions of Tantra and Dzogchen.

According to Tibetan Buddhism, Dzogchen is the natural condition of the Mind and ultimate basis of all sentient Beings, experienced as pure, all-encompassing, timeless awareness. This sentient awareness has no form of its own and yet is capable of perceiving, experiencing and expressing all forms of Being. It can do so without being affected by those forms in any way. This pure awareness of the nature of Being is called *rigpa*. Each Element is analogous to experiential sensations of reality. In Bön, the Elemental processes are also fundamental metaphors for working with external, internal and secret energetic forces.

Aristotle and the traits of the Elements

Aristotle saw the Elements as combinations of 2 sets of opposite qualities: hot and cold, wet and dry. Even though his view serves no practical purpose when considering the physical structure of Being, it is interesting when represented as a Square diagram, thus emphasizing the interaction of Dual qualities.

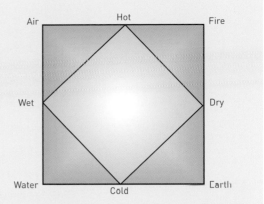

	Hot	Cold	Wet	Dry
Earth		•		•
Water		•	•	
Air	•		•	
Fire	•			•
Ether				

Hot   Expansion
Cold Contraction
Wet   Dissolution
Dry   Crystallization

Native American Elements

Native Americans recognize earth, air, water and fire Elements. Their system is similar to that of the Taoists in that they assign these Elements to the 4 cardinal points (the Cross) on the Medicine Wheel. Each cardinal point has one of the 4 seasons, depending upon their characteristics:

Fire	East	Spirit and intuition
Earth	West	The physical body and senses, physical things of the Earth
Air	North	The Mind, consciousness and freedom to move wherever you wish
Water	South	The soul and our emotions

The Medicine Wheel represents the perfect balance of the active and passive energies that complement each other. So, as for Taoists, the directions and seasons represent attributes and directions of the individual in the course of their life.

Left **Native American women carrying out a corn ritual. They are standing in an open Circle around ceremonial gifts, comprising the Elements, given to each of the Directions.**

12 Classical Elements and the Platonic solids

Science is organized knowledge. Wisdom is organized life.

IMMANUEL KANT, GERMAN PHILOSOPHER
(1724–1804)

Above **17th-century mathematician and astronomer Johannes Kepler was convinced that his studies revealed God's geometrical plan for the Universe, as shown in his *Harmonices Mundi*, 1619.**

Good order is the foundation of all things.

EDMUND BURKE,
IRISH STATESMAN
(1729–1797)

PLATO (424–347 BC) WAS AN INITIATE OF THE IONIAN, or Milesian, School of Greek philosophy (6th century BC). He devised a theory that the basic structure of matter evolves from a simple geometric shape to form more complex geometric shapes. He believed that the entire Universe could be explained in mathematical terms and that we could master the Universe if we discovered the numbers hidden in all things. His school also believed, in a similar way to Taoism, that the world was in a constant state of flux and that a union of opposite principles produces most objects – objects that change and merge according to the rule of balance and proportion found in geometry.

The Platonic solids

Plato later conceived of matter as consisting of atoms with the geometrical shapes of 4 of the 5 regular geometrical solids, as described by Plato in the *Timaeus*. These are called the "Platonic solids" and modern geometry stems from these forms. Many believe Plato acquired his information from Pythagoras, who in turn stated that he received much of his information from Egypt, where he spent some 20 years being initiated in the ancient Mystery Schools of the Egyptians. Archaeologists have found some of these geometric shapes carved in stones dating back some 20,000 years. Even before this era we find the same shapes incorporated into the writings of even older civilizations.

The 5 solids

Firstly it is worth noting that there are 5 Platonic solids. Again the number 5 features in relation to the creative process and facilitation of forms of Being. It is interesting that all the surfaces of the Platonic solids consist entirely of Triangles, Squares and 5-sided pentagons.

The 5 Platonic solids include the tetrahedron, the cube (also known as the "hexahedron"), the octahedron, the dodecahedron and the icosahedron. These are the only 5 regular polyhedrals (the only 5 solids made from the same

Geometric crystals

Platonic solids are, of course, not the true shapes of atoms, but it turns out that they are some of the actual shapes of packed atoms and molecules, particularly visible in crystals. The mineral salt (halite, NaCl) occurs in cubic crystals; fluorite (calcium fluoride, CaF_2) in octahedrons; and pyrite ("Fool's Gold", iron sulphide, FeS_2) in cubes and dodecahedrons.

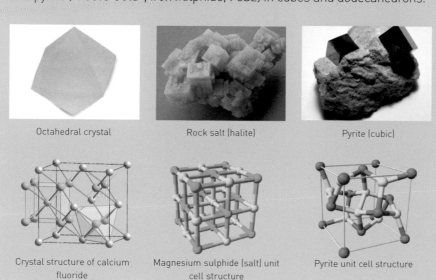

Octahedral crystal   Rock salt (halite)   Pyrite (cubic)

Crystal structure of calcium fluoride   Magnesium sulphide (salt) unit cell structure   Pyrite unit cell structure

> The marble not yet carved can hold the form of every thought the greatest artist has.
>
> MICHELANGELO,
> ITALIAN ARTIST
> (1475–1564)

equilateral, equiangular polygons). Each of these forms conforms to certain rules, as listed below. They have only:

- one edge length
- one face size
- one angle throughout
- all the angled points fit on the surface of a sphere, the shape considered to represent the Absolute Void.

The cube and octahedron are Dual to each other, meaning that one can be created by connecting the mid-points of the faces of the other. The icosahedron and dodecahedron are also Duals as 2 mutually perpendicular, mutually bisecting golden rectangles can be drawn, connecting their vertices and mid-points respectively. A tetrahedron is a Dual to itself.

Classical Elements and the Platonic solids

The Greek Timeus of Locri linked the 5 Platonic solids and the 5 Elements. The Sphere was used in alchemy to represent the Circular Vessel of the Void; that which contains all things and within which the Platonic solids can fit. A smaller sphere can be inscribed within the voids. This can be illustrated in two dimensions using a Square and Triangle. The Platonic solids and their Elemental natures are inherent in all manifestations of Being, as will be shown in this chapter.

The Platonic Solids and their Duals

These simple, aesthetically pleasing regular solids and their Duals have been a key part of the study of geometry for thousands of years.

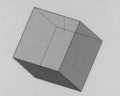 

Cube
6 Squares

Tetrahedron
4 faces are identical
equilateral Triangles

Octahedron
8 Triangles

Icosahedron
20 equilateral Triangles

Dodecahedron
12 pentagons

Platonic solid Duals

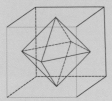  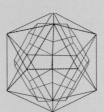

Cube and octahedron

Icosahedron and dodecahedron

A tetrahedron is a Dual
to itself

> Each thing is of like form from everlasting and comes round again in its cycle.
>
> MARCUS AURELIUS, ROMAN EMPEROR (AD 121– 180)

Summary table – Platonic Solids, Duals and Elements

Shape	Element	Faces	Dual
Tetrahedron	Fire	4	Tetrahedron
Cube	Earth	6	Octahedron
Octahedron	Air	8	Cube
Dodecahedron	Water	12	Icosahedron
Icosahedron	Ether	20	Dodecahedron

Astrological Elements and destiny

Astrological Elements are organized into the triplicities that are 4 groups of 3 signs represented by air, earth, fire and water. Signs that belong in the same aspect belong in the same trine, for example the water signs are Pisces, Cancer and Scorpio. This links 4 earthly foundations and physical life with our stellar destiny – as drawn out by the movements of the heavenly bodies.

Above **Kepler's nested Platonic solids within the sphere and model of the Solar System,** from *Mysterium Cosmographicum* (1596).

Sacred Greek architecture

In Ancient Greece the cube symbolized earthly foundations. The Golden Section, or Phi Ratio, symbolized wisdom and is itself a "divine" proportion that provided a unique division of space. Making use of these attributes, a god of the Earth would have a temple based on cubic geometry and a god in the heavens would have one using Golden Section proportions.

Above **Greek astronomer Hipparchos** (c190–c120 BC) **shown in a wood engraving** c1880.

Left above **The alchemical symbols for earth, water, air and fire.**

Left below **The four elements, depicted in** *Viridarium Chymicum* (1624) **by D. Stolcius von Stolcenberg.**

Interpretation of form

The Platonic solids, Classical Elements and the importance of numbers are to be interpreted metaphysically as well as literally. Pythagoras stressed the significance of form rather than the structure of matter, where the objects of the physical world are merely reflections of ideas. Only the changeless eternal forms can be the objects of true knowledge. Certainly Plato believed that the Platonic solids were shadows, or representations, of reality. According to Parmenides, the leader of the Eleatic school of philosophy, the appearance of movement and the existence of separate objects are merely illusions, a philosophical view now supported by modern science (see page 102).

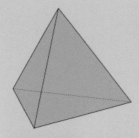

Top and above
Tetrahedron – fire: our ability to make and control fire at will was a major evolutionary development.

Platonic solids and their Elemental natures

The Elements and Platonic solids constitute the simplest essential parts and principles of which Being is comprised. Inspired by the observations of the changing states of matter and the role of heat it was believed that as water descends (as in rain) and fire ascends (as heat does), air and earth are formed.

FIRE IN THE MIDST

According to Ancient Mystery School teachings the tetrahedron was regarded as one of their most magical and powerful symbols. "Pyramid" means "fire in the midst" and because of this the pyramidal tetrahedron represented the Element fire. Each face of the 6-pointed tetrahedron star, also known as the Star of David, represents different forms of alchemical fire. The 3 upper Triangles represent solar fire, volcanic fire and astral fire. They are also linked to the 3 astrological fire signs Aries, Sagittarius and Leo. The lower 3 Triangles represent latent heat. The tetrahedron, a 3-dimensional Triangle, was also used by alchemists to represent the union of the 4 Elements that facilitate the creation of physical life that was then animated by the fifth Element, ether, or the Life Force.

DESTINY OF MATTER WITH SPIRIT

The octahedron represents the Element air and comprises 2 opposite Square-based pyramids of Heaven and Earth, each being a reflection of the other. Octahedrons have 8 faces and 12 edges. The 12 edges represent the signs of the zodiac and link destiny of Matter with spirit. Its Dual is the cube, the solid within which it can be found in the Grid of Space and delineates the directions in space. The link to astrology comes from the Square medieval ideogram used to represent the 4 physical Elements in their combined form of astrology, philosophy and

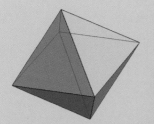

Above and above right **Octahedron – air: we even manipulate the air for our own requirements, as for hot air or helium balloons.**

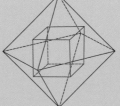

Octahedron in a cube

Medieval ideogram for the four Elements

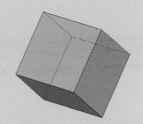

alchemy. Elements are placed in Triangles in the corners of a Square. This figure is shown on the facing page (bottom right), and it is a symbol that can be found within the Grid of Space and the 8-fold division of Space-Time. 2 Elements are allocated to each of the Earthly and Heavenly aspects of Being. As shown on page 95, at the top are the Triangles representing the ethereal Elements of air and fire; at the bottom are the Triangles representing the material Elements earth and water.

SOLID FOUNDATIONS

Naturally, since the Square is associated with being grounded and the Earth, the cube represents the Element earth, since cubes have Square faces and are solid. Surprisingly, models of cubes are fragile and you rarely find examples of cubes in Nature. However, once a tetrahedron is inside a cube, so that its edges lie on the faces of the cube and the vertices meet 4 corners of the cube, it is stabilized. The 6 faces of a cube are associated with the planets: the Moon, Mercury, Venus, Mars, Jupiter, Saturn and the Sun is in the Centre.

Top and above **Cube – earth. Solid materials have a multitude of roles, for example in construction.**

GOLDEN WATER

Water is represented by the icosahedron, comprising 2 pentagonal (5 sides) caps on top and bottom and 10 Triangles around the middle. Every single line is based on the Phi Ratio, even the crossing patterns. Additionally, every line that is connected is a Phi Ratio Triangle.

THE LIFE FORCE IN BEING

The dodecahedron represents the Life Force, Chi, ether, or quintessence. It has a hexagonal (6) outline and a pentagonal (5) outline when viewed from opposite sides.

Above **Dodecahedron – the Life Force: an intangible force that is behind growth and decay, expansion and contraction.**

Top and above **Icosahedron – water: we seek out pure water, liquid that is key to our survival.**

13 What is matter?

Nature is the thriftiest thing in the world; she never wastes anything; she undergoes change, but there is no annihilation, the essence remains – matter is eternal.

HORACE BINNEY, AMERICAN LAWYER (1780–1875)

> Everything we call real is made of things that cannot be regarded as real.
>
> NIELS BOHR, DANISH PHYSICIST (1885–1962)

THE WORD "MATTER" COMES FROM THE SANSKRIT WORD *matr* "to measure". It is also the source word for "matrix", "metre" and "Maya". Physical reality appears to be measurable, made possible by the Matrix of Space-Time-Being, but actually reality is Maya created by our senses. Maya is the Eastern concept that all form, though manifest, is an illusion veiling true reality. Behind Maya, supporting it and allowing it to exist, is the wider reality of Brahman, which is all energy unmanifest.

Chemical Elements

Modern chemistry (chemical Elements, chemical compounds and mixtures) has evolved from medieval, Islamic and Greek models of the Classical Elements. Modern-day thinking about physical matter changed when it was discovered that water decomposed into the molecules of hydrogen and oxygen when an electrical

---

Stable chemical elements and the number 9, perfection
- $9 \times 9 = 81$ = number of stable elements.
- Only 81 stable chemical elements exist. This excludes chemical elements that can only be artificially synthesized.
- $81 = 3 \times 3 \times 3 \times 3 = 3$ to the power of $4 = 9 \times 9$
- Also $1/81 = 0.012345679012345678901....$ a fraction counting upward with the number 8 missing.

From his observations, German scientist Peter Plichta in *The Cosmic Cross* worked out that the 81 elements and their atomic numbers are reciprocally linked to each other and that Nature appears to be organized using the decimal systems (just as we have ten fingers). Plichta regards the Moon and Earth as Dual planets; the Moon influences the movement of water and the relationship between the mass of the Earth and the Moon is 1:81.

current was passed through it. Then we became obsessed with discovering the smallest unit of matter possible. Molecules were discovered to be collections of atoms and these were given the label "chemical elements". In turn, it was found that atoms consist of collections of "elementary particles", i.e. protons, neutrons and electrons, neutrinos and muons. The amount of energy required to break the subatomic forces binding these miniscule particles together is phenomenal.

The Crab Nebula

One of the largest mosaic images taken by NASA's Hubble Space Telescope is a six-light-year-wide expanding remnant of a star's supernova explosion that took place in AD 1054. The colours in the image indicate the different elements that were thrown out during the explosion. Blue strands in the outer part of the nebula represents neutral oxygen; green is singly ionized sulfur; orange is mainly hydrogen (the remains of the star) and red indicates doubly ionized oxygen. Note the fractal patterns of the electron trails (see right).

The inner blue light comes from electrons spinning at nearly the speed of light around magnetic field lines from the neutron star gyrating at its centre. Much like a rotating beam in a lighthouse, the star emits twin beams of radiation that seem to pulse at 30 times a second.

The Crab Nebula (NASA)

The nature of matter at an atomic level

These drawings are schematic representations of the orbital electrons in five different atoms. Note that they have a centre and rings of orbiting electrons, like the Holon.

Hydrogen, the first, is the most common element in the nucleus, with a single electron orbiting a nucleus.

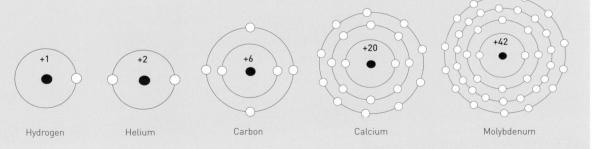

| Hydrogen | Helium | Carbon | Calcium | Molybdenum |

+1   +2   +6   +20   +42

Laying out the chemical elements

The Periodic Table of Elements (see immediately below), is the most well-known organization of the elements. But when laid out in Circular or Spiral formats, their natural grouping according to relationships based on their underlying atomic structure is more obvious.

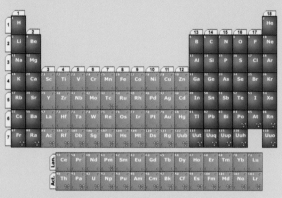

The Periodic Table of Elements in its traditional layout

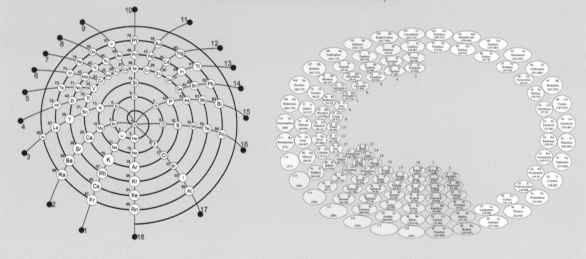

The Periodic Table of Elements as a Spiral
(by Jan Scholten)

The Periodic Table of Elements as a Circle

> The mind of man has perplexed itself with many hard questions. Is space infinite, and in what sense? Is the material world infinite in extent, and are all places within that extent equally full of matter? Do atoms exist or is matter infinitely divisible?
>
> JAMES C. MAXWELL,
> SCOTTISH PHYSICIST AND MATHEMATICIAN (1831–1879)

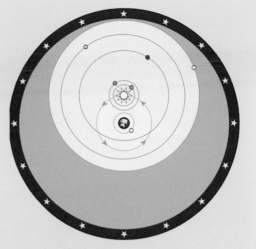
The basics of the Tychonian geocentric system

Tycho Brahe (late 16th century) merged the mathematics of the Copernican system with the philosophical Ptolemaic system. His motions of the planets resemble the inner workings of a mechanical clock in the geocentric system.

The Moon, Sun and fixed stars revolve around the central Earth and are shown with blue orbits. The objects on orange orbits, namely Mercury, Venus, Mars, Jupiter and Saturn, revolve around the Sun. All the planets are encircled by a sphere of stars that are fixed only with respect to each other, since the sphere revolves around the Earth.

Above **On a planetary scale, as part of our solar system, the planets orbit the Central Sun.**

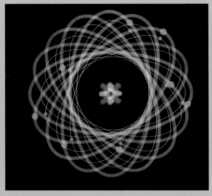

Far left **Schematic representation of the planets and the asteroid belt in our solar system (not shown to scale).**

Left **Examples of typical patterns made by the electrons around a nucleus.**

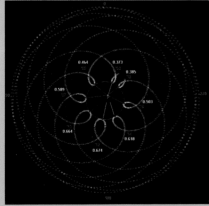

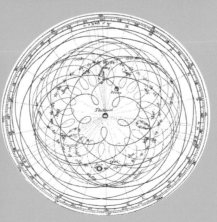

Far left **The motion of Mars relative to the Earth (central blue dot), showing opposition and comprising three-day steps, with the oppositions listed by year and their closest distance.**

Left **The path of Mercury's orbit for 7 years and Venus for 8, in which time the passage of Venus returns to almost the same apparent position.**

Carbon and life as we know it

With respect to the creation of Being the importance of the Element carbon (denoted with a C and having an atomic number 6) cannot be overstated since it occurs in all known organic life. Without it physical organic life, as we know it, could not exist.

Carbon has a unique ability for forming multiple bonds and chains. Every carbon atom wants to bond with as many as 4 other carbon atoms. This facilitates the creation of long chains and rings, sometimes as a compound when it joins together with other elements' atoms (almost always hydrogen, often oxygen, sometimes nitrogen, sulphur or halides). Chains and rings are fundamental to the way carbon-based life forms build themselves. Note that a chain is a line and the line is a force between two opposing Duals, like north and south, while a ring is a Circle. At one extreme, carbon atoms connect in a configuration to make diamond, the hardest substance, and at the other extreme they combine to make soft graphite.

Most of the compounds known to science are carbon compounds, which are often called organic compounds because it was in the field of biochemistry that they were first studied. The compounds carbon forms with metals are generally considered to be inorganic.

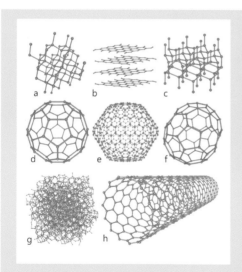

Molecular configurations of carbon – the basis of "life"

The 8 different molecular configurations that pure carbon can take are:
a) Diamond
b) Graphite
c) Lonsdaleite
d) to f) Fullerenes C_{60}, C_{540}, C_{70}
g) Amorphous carbon
h) Single-walled carbon nanotube

Left **8 allotropes of carbon (by Michael Ströck, February 7, 2006).**

Polymorphism

Polymorphism is the ability of a solid to exist in more than one crystal form. For pure chemical elements, polymorphism is known as "allotropy". For example diamond, graphite and fullerenes are different allotropes of carbon.

Tetrahedron in chemistry

The tetrahedron is the only three-dimensional shape whose vertices are the same distance from each other. Apart from the sphere no other solids have fewer than 4 faces or 4 vertices. While the sphere encloses the most volume capable of holding all the solid shapes, space within the tetrahedron is minimal. This makes the tetrahedron the strongest and most stable solid as its root shape is the Triangle.

Tetrahedrons are common in organic and inorganic chemistry. Many elementary molecules use it as a frame, for example CH_4 (methane), C_2H_6 (ethane) and amino acids. In each of these three molecules, a carbon or nitrogen atom sits at the centre of the tetrahedron at whose 4 corners reside 4 smaller hydrogen atoms.

Diamond is the hardest natural substance and the best conductor of heat. Diamond atoms are structured as a tetrahedral network of carbon atoms and energy bonds that are short and strong.

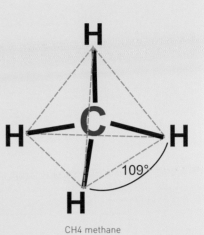

CH4 methane

Underlying order in solids

Crystalline solids, such as salt, diamond and quartz, have a molecular structure with a specific geometric shape. At a given melting point all the molecules begin to break free of their bonds. These solids typically form by cooling a liquid slowly, giving the molecules time to arrange themselves in a geometric crystalline structure as their bonds form. They do this in the densest shape possible using a repeated pattern called a "unit cell". Usually, two different patterns occur; hexagonal (hexagon and hence Triangular basis) and face-centered cubic (Square basis).

Amorphous solids such as coal and glass have a molecular structure with no specific geometric shape. As the amount of energy needed to break the bonds varies from molecule to molecule, so does the temperature at which the solids melt. Amorphous solids form when a liquid is cooled quickly so that there is no time for the molecules to arrange themselves into a crystal.

Because of their underlying geometry crystalline solids are harder and firmer than amorphous solids. Coal and diamond are both made from elemental carbon, but coal is amorphous and diamond is crystalline. Because of its crystal structure diamond is very hard, whereas coal can be scratched easily. Diamond's sparkling transparency is due to its crystalline basis and coal is black and dim because the disorder of its atoms does not allow light to pass through it.

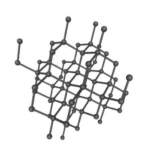

Structure of diamond

Cut diamonds

"Life exists in the Universe only because the carbon atom possesses certain exceptional properties."

SIR JAMES HOPWOOD JEANS, ENGLISH PHYSICIST (1877–1946)

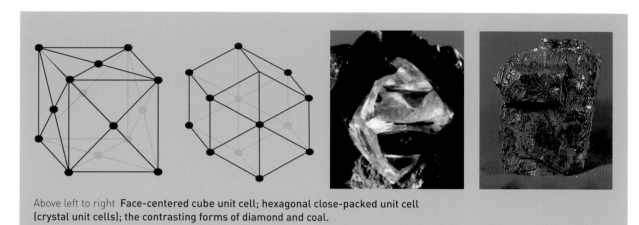

Above left to right **Face-centered cube unit cell; hexagonal close-packed unit cell (crystal unit cells); the contrasting forms of diamond and coal.**

All is flux, nothing stays still.

HERACLITUS OF
EPHESUS,
ANCIENT GREEK
PHILOSOPHER
(c535 BC–c475 BC)

Perpetual transmutation

At last modern scientific understanding of matter is supporting the view that matter is not made of isolated blocks constructed in hierarchies; recent advances in physics have shown that it is always in flux. At the subatomic level, matter does not exist with absolute certainty in definite places, but can only be said to have "tendencies" to exist. All particles can be transmuted into other particles; created from energy and vanishing into energy in a continual process, like an endless sea of potential involved in a complex dance. Each particle helps to generate every other particle, which in turn generates it.

Everything exists in a state of flux at the subatomic level and even up to the world of Nature visible to our naked eye in the states of matter. Nature's forms are dying, growing, moving and changing from one state to another continually. It is not possible to stay still and just "be", as everything perpetually becomes something else, moved along by the combined forces of the Circle of time and Spiral

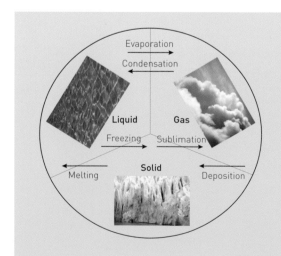

States of matter

Classical Element water corresponds to the liquid state, air to the gaseous state, earth to the solid state and fire represents temperature. Temperature is the feature of chemical processes that facilitates changes in matter from one state to another, since it causes chemical reactions and the reorganization of atoms to create new forms.

Left **The states of water shown as a cycle within a Circle.**

Life Force. Nothing can hold back these forces or move against them. They are a necessary part of growth, learning, evolution and transformation of each individual Being and the sum of all things.

Phases and states of matter

A "phase" is a process of change of condition from one state to another, but is often used as a synonym for a state of matter. Phase boundaries relate to changes in the organization of matter, such as a change from liquid to solid or a more subtle change from one crystal structure to another. Here a phase is a set of equilibrium states bounded by variables such as pressure and temperature. Many compositions will form a uniform single phase, but depending on the temperature and pressure even a single substance may separate into two or more distinct phases. The state is the condition of the matter within the phase boundary.

Even with our knowledge of more than a hundred types of atom, modern science can only find 4 states of matter: solid, liquid, gas and plasma, or electronic incandescence (gases that "burn" in the Sun and stars and cause the glow in fluorescent lights). Distinct phases may also exist within a given state of matter.

Phases in water

At a critical point, water as liquid and gas becomes indistinguishable. This unusual feature is related to ice having a lower density than liquid water. Increasing the pressure drives water into the higher-density phase, which causes melting. Another interesting feature of the water phase diagram (see below right) is the point at which the solid-liquid phase line meets the liquid-gas phase line. The intersection is referred to as the "triple point". At the triple point, all 3 phases in water can coexist.

Water is a well-known example of a material with several distinct solid states capable of forming separate phases. For example, water ice is ordinarily found in the hexagonal form Ice Ih, but it can also exist as the cubic ice Ic, the rhombohedral ice II and many other forms.

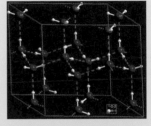

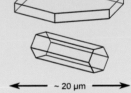

← ~ 20 µm →

Above **Hexagonal Ice Ih Molecular structure and crystal form**

Far left **Ice crystals.**

Left **A phase diagram of water.** The yellow dot is a triple point; the red dot is a critical point; the deep blue area is water; grey is vapour; white is ice and the green line is supercooled water.

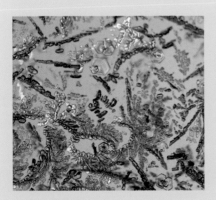

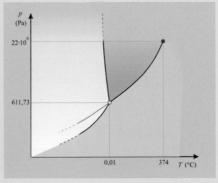

Above **At each extreme of scale, cosmic and subatomic, there are primarily vast tracts of empty space.**

Defining things by what they are not

Underlying the latest theories of matter at a subatomic level is the major discovery of the early 19th century, that the underlying reality of the Universe is based on fields rather than building blocks. Fields are a condition in space that has the potential to produce a force, a potential that is realized by forces interacting with each other. The potential is the dynamic of each shape, which we intuitively know and harness using our Mind's intent in symbols and structures based on geometry.

Now scientists are debating whether the nature of fields may be based on a phenomenon beyond that of forces alone. After the discovery of fields came the discovery of relativity, linking space and time. Energy moves in these fields, through the impenetrable, communing web that binds Space-Time-Being together. The unifying concept of fields, rather than of hierarchies, is one that is moving into many spheres of study, ecological, social and biological, as a more reasonable explanation for the way life organizes and interrelates.

Things are defined by what they are not. This means that they have to be described by their boundaries, where they become something else, such as the boundaries where earth becomes air. Look around you right now. There are no lines separating the edge of the chair from the floor or your body. It is their colour, texture, tones and form, or the changes in the fields that make it appear that different things are there. From this perspective, the emphasis is placed upon relationships, so that the underlying geometry becomes more important than the actual physical being. Zoologist D'Arcy Wentworth Thompson (1860–1948) believed that a form is shaped by the physical forces acting upon it rather than by genetic coding in the nucleus of the cell.

Pressure of the forces

Like temperature, pressure can break atomic bonds to change the state of a material and hence form. Physicists have identified 4 natural forces – gravity, electromagnetism, the weak radioactive force and the strong nuclear force.

On the surface the forces appear to be distinct, but as their effects are monitored at ever-increasing temperatures their differences vanish and matter is degraded into less and less useful forms. It takes a very long time to turn these back into useful forms. A good example of this is the process of releasing energy by burning fossil fuels that have taken thousands of years to create. Most physicists believe that it will eventually be shown that these forces are different manifestations of just one basic force. Pure simplicity is only visible with extreme heat and one force.

> Nature uses as little as possible of anything.
>
> JOHANNES KEPLER,
> GERMAN ASTRONOMER
> (1571–1630)

Forms and forces

The Matrix of Space-Time-Being can be compared with the theory of fields and forces. Since there are 4 forces, intuitively they must be linked to the Square and the emergence of a form of Being in Space-Time via the Point. 4 = Square and the sets of Duals that establish boundaries and stability. From its central Point a Being is supported and sustained energetically by the opposing forces of the Directions.

Newtonian theories relied on a constant world where energy and matter interacted on a strictly causal basis. Field theory added the concept of gravity with the discovery of the force of electromagnetism. It shows how matter is held together in various forms between the opposing grips of the forces. Planets, people, plants and animals are all composed of closely packed arrays, or lattices, of atoms. These arrays are disrupted by the pressure of forces, which crush and stretch it to alter the form. All these forms have a centre of gravity, just as humans do, near the navel.

Look at any solid object. The appearance of its surface is due to the rapidly oscillating atoms. They move so fast that to our nervous system they give the impression of a smooth surface through our sense of touch and sight. The same is true for hearing as sound is made of discrete units moving so fast that they become a hum. All of our senses are fooled into believing that a solid reality is there and solid by rapid-cycle vibration so that each sensory experience appears to be continuous.

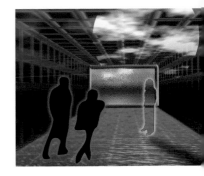

Above **Invisible forces of the Matrix hold, mould and sustain reality.**

Form and geometry

In sacred geometry every shape used symbolically, in whatever application, has a function to hold, or sustain, energy. Our human creations in art and architecture provide imagined boundaries for the unbounded and frameworks for the hugely complex, mirroring the process of creation. At the end of all the analysis there is no matter, only form, and form is facilitated by the magic of geometry. Apparently solid forms comprise patterns and shapes that are crystallizations, or "intangible concentrations", of vibrating waves made of energy that flows, swells, bubbles and whirls with eddies and currents (as exhibited in the metaphors of water and air Elements). Everything in our Universe is a feature of this energy, forming within the Matrix of Space-Time-Being and vibrating down the Planes into this reality. Different concentrations of structural patterns of waves unite to form the myriad chemical elements that react with one another to form physical substances. Matter only appears to be solid because we are constituted of similar wave forms, resonating within a clearly defined range of frequencies that control the physical processes of our world.

"When it comes to atoms, language can be used only as in poetry. The poet, too, is not nearly so concerned with describing facts as with creating images."

NIELS BOHR,
DANISH PHYSICIST (1885–1962)

14 Geometry in multiple dimensions

Our soul is cast into a body, where it finds number, time, dimension. Thereupon it reasons, and calls this nature necessity, and can believe nothing else.

BLAISE PASCAL, FRENCH MATHEMATICIAN, PHYSICIST, WRITER AND PHILOSOPHER (1623–1662)

THIS CHAPTER OUTLINES SOME VERY ABSTRACT IDEAS AND MATHEMATICS to do with dimensions of reality beyond the boundaries of our physical existence and senses. Overuse of the word "dimension" in physics and mathematics and conjectures about alternate Universes, the spirit realms and fantastical possibilities, only clouds communication and comprehension of the concepts of time, space and existence of Being. By definition, a dimension is a "measure" of some kind, like width, breadth and height. These measures relate to the mathematical dimension of space, or an object, which is defined by the minimum number of co-ordinates needed to determine a Point within it. So "dimensions", or measures, are clearly pertinent to our understanding the nature of space and structures in space, but how are multiple measurements of time and Being to be more clearly described?

Above **Even the imagined "other worlds" of artists and thinkers utilize the key principles of geometry.**

Aspects of Reality

There are 3 Aspects of reality: space, time and Being. They originate from the seed Central Point (1,I) that is contained by the Absolute Void (0) yielding infinite possibilities to be everything, anywhere, anytime. In this Centre is the eternal, ever-present Moment in Time, the place Here in space and the all-seeing I/ Eye of Being. At the Centre, from our view of reality, we can know ourselves in this reality as "Now, Here, I am". Within the Gateway to the Heavens model is the Matrix of Space-Time-Being made from the Grids of Space (Square), Grid of Life (Triangle) and Grid of Time (Circles). As a blueprint for the Universe it is visible in every feature of Nature working on minute and immense proportions and it spawns tremendous variety. Dimensionality is therefore obviously necessary as it builds up reality as we know it and also beyond what we can perceive. To fully understand dimensionality we need to explore the different ways of measuring reality and then how these measures are combined.

Mutually dependent measurements of the Aspects

Dimensions, or measures, of space are the most natural to us; the Point has 0 spatial dimensions (0D), a line has one dimension (1D), a plane two dimensions (2D) and a solid body has three measurable dimensions (3D). Time is also measured, but these are measurements of something far less tangible to us than space. Time measurements of a second, minute, hour, day, season, years, eons and ultimately eternity, turn around the Moment. Only the singular Moment captures an instant in time. Dimensions of Being relate to the definition and scope of what it is "to be" and the perception of identity, or "I"; or the measures of sentience (see page 115).

Since they originated from the same source, the subsequent dimensions, or measurements, of the 3 Aspects of time, space and being are mutually dependent. Consider how the Trinity of the human body, soul and Mind can each be studied, but never isolated. Similarly each Aspect can be analyzed logically, but it is vital to remember that their context as part of the Whole is not so logically comprehended. To do this you need to rely more on your intuition.

> *"Once Ptolemy and Plato, yesterday Newton, today Einstein, and tomorrow new faiths, new beliefs and new dimensions."*
>
> ALBERT CLAUDE, BELGIAN BIOLOGIST (1899–1983)

> *"The key to growth is the introduction of higher dimensions of consciousness into our awareness."*
>
> LAO-TZU, CHINESE PHILOSOPHER (6TH CENTURY BC)

Measurement of time

The ultimate cycle of eternity surrounds the indivisible Moment. Radial Circles surround the Moment within eternity as a Holon. There are no gaps between these cycles, which rotate around the Moment and create the perception of past and future. Time cannot be given traditional spatial measurements of width, length and depth. In Einstein's model of Space-Time (see page 118) time is a "dimension" in its own right.

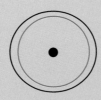

| Eternity around the Moment | Smaller cycle (e.g. a second) | Larger cycle (e.g. an eon) | Grid of Time (summarizes the impenetrable radial Circles of the Holon) | Above **In the clock, arms rotate around the Moment and circular cogs are used in the mechanics of measuring time.** |

Measurement of the Aspect of Space

Spatial dimensions are built up through a sequence of doubling, squaring and cubing.

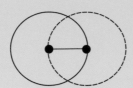

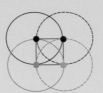

Point 0 dimensions (0D)	Line 1 dimension (1D) 1 unit of measurement	Square plane 2 dimensions (2D) 1 x 1 = 1 unit squared	Cube 3 dimensions (3D) 1 x 1 x 1 = 1 unit cubed	2 units cubed 2 x 2 x 2 = 8

Within the first 10 numbers only 1 (1 x 1 x 1 = 1) and 8 (2 x 2 x 2 = 8) are cubic numbers and their root numbers of 1 and 2 are parents of the other numbers.

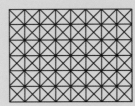

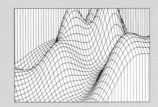

Above **Note that a 3-cube viewed directly from a corner, centred on its origin, looks like a hexagram.**

Above **Grid of Space: summarizes the 8-fold nature of 3-dimensional space.**

Above **The 3-dimensional spatial Grid of Space enables the perception of distance and hence volume in forms of Being.**

Measurement of the Aspect of Being

The latent potential for different types of Being permeates the Void. As seen in Chapter 13, energetic Beings manifest by vibrating at different degrees, becoming denser the slower the vibrations are. The Grid of Life based on the Triangle is influenced by the cubing in the Grid of Space. Cubing enables physical forms of Being to assume a solid "body" with spatial measurements.

2-dimensional plane hexagram

3-dimensional hexagram or "star tetrahedron", in a 3-dimensional space enabled by cubing (note how it fits neatly into a cube).

Measurement of the Aspect of Being (continued)

However, measurement of Being is not to do with the shape and size of the physical, or for that matter ethereal, body. Measurement of Being is defined by the scope of sentient awareness of the Being, as described in Chapter 10 (see page 79). Just as time is an intangible concept, so is sentience and it is not possible to give sentience spatial measurements, such as width and height. Instead an Eye is used for describing degrees, or measures, of sentience symbolically. The more open and the bigger the Eye the more sentient the Being is.

The original Eye is "all seeing" pure consciousness, represented here as a "giant" eyeball comprising the Circle with central Point and Triangle within the Circle. This Eye is completely open and "is" all parts of the Triangle. It is completely sentient of past/present/future, mind/body/spirit and all other Trinities.

An Eye within a Triangle represents a degree of consciousness "frozen" in a physical form, just as we are. The Eye appears to have no specific location in the Triangle and this is because the Triangle is like a window for the Eye to look through moderately, or completely. This window outlines the "view" available to the physical Being.

First-Level Rhombus
2 Triangles within the vesica piscis

Third-Level Rhombus

When the Third-Level Rhombus is seen within the context of the extended Grid of Life (outlined in orange) it sits neatly within a large hexagram. The hexagram is the balance and union of the original 2 Triangles in the First-Level Rhombus. When the Eye is shown in this symbol it can now be seen to be located at the Centre of the hexagram (and Triangles) due to the Grid of Life. This unification of opposite Triangles is to remind us that Being has 2 parts of matter and spirit bridged by the Mind, which is our Eye perceiving reality. Physically and spiritually we can close, partially open or totally open our Eye of sentience by opening the Triangle, or trinity, of our two physical eyes and our Third Eye of insight to allow in more information about reality.

As the same fire assumes different shapes when it consumes objects differing in shape, so does the one Self take the shape of every creature in whom he is present.

Marcus Aurelius,
Roman Emperor
(AD 121–180)

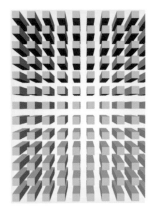

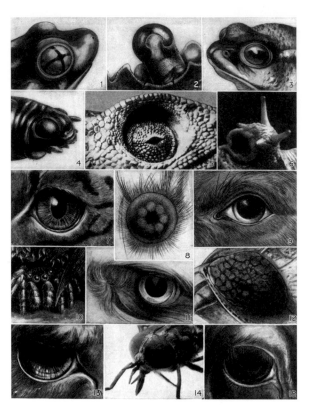

Above If you look directly at any solids, such as a cube, you see only its flat face, which is how a cube appears in the 2-dimensional Grid of Space. By moving your Eye, Here, Now, to look at solid Beings from different perspectives and scales in Space-Time, a sense of volume and distance is created in your Mind.

Left According to some beliefs all Beings manifest in reality have a degree of consciousness, but how sentient each and every form of Being is, be it a planet, pebble, fire, animal or plant, it is not for us to judge, as we have not materialized in their form. Certainly every animal with eyes will see reality from a different perspective to our human one.

SUMMARIZING THE ASPECTS, THEIR LOCATORS, SHAPES AND MEASURES

Absolute	Aspect	Locators	Shape		Measures
Eternity	Time	Moment	Circle	●	Second, minute, hour, day, month, year
Infinity	Space	Point	Square	■	Line, area, volume
Existence (energy/consciousness)	Being	Eye	Triangle	▲	Sentience, degree of consciousness

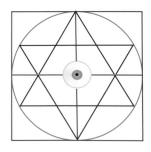

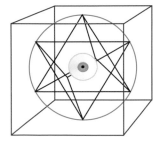

Left The 3 Aspects of Reality combined in the 2-dimensional and 3-dimensional Matrix of Space-Time-Being embody all dimensions of reality. At its Centre it is experienced as "Now, Here, I am".

Euclidean and non-Euclidean geometry

The Greek mathematician Euclid's *Elements* is the earliest-known systematic discussion of geometry. Plane geometry and solid geometry are well known. In non-Euclidean geometry shapes and structures do not map directly to n-dimensional Euclidean system (where n denotes a number from 1, 2, 3 upward). The main difference between Euclidean and non-Euclidean geometry is the dynamics of what happens to parallel lines.

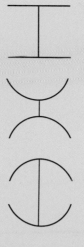

Top left Euclidean geometry – the parallel lines remain at a constant distance from each other, even if they are extended infinitely.

Centre left Hyperbolic geometry – the lines curve away from each other, increasing in distance. These lines are often called "ultraparallels".

Bottom left Elliptic geometry – the lines curve toward each other and eventually intersect.

Above and above right **Euclidean geometry in a grill matches our perception as we look down an avenue of trees.**

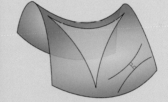

Hyperbolic Triangle

Hyperbolic icosahedron

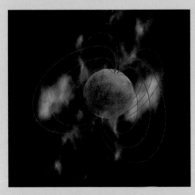
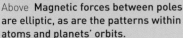

Above **Magnetic forces between poles are elliptic, as are the patterns within atoms and planets' orbits.**

Above right **Stunning elliptic and hyperbolic geometry in microscopic radiolaria.**

Right **Elliptic arches within a passageway create a different visual experience to that of straight lines.**

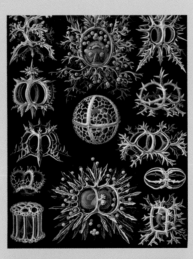

$\Omega_0 > 1$

$\Omega_0 < 1$

$\Omega_0 = 1$

Above **Theories about the shape of the Universe correspond to elliptic, hyperbolic and Euclidean geometry. Open (hyperbolic) and flat (Euclidean) Universes are infinite and by travelling in a constant direction you will never get to the same point. The closed Universe (elliptic) is of finite size and travelling far enough in one direction will return you to your starting point.**

> I existed from all eternity and, behold, I am here; and I shall exist till the end of time, for my being has no end.
>
> Kahlil Gibran,
> Lebanese-American
> artist, poet and writer
> (1883–1930)

Einstein's "fourth dimension" and relativity

To the traditional 3 measurable dimensions in Euclidean space (line 1D, plane 2D and volume 3D) Einstein added time as a further fourth dimension and defined the Space-Time continuum. Time acts with space, not independently from it and he proved that the aspects of time and space are two features of the same thing. Einstein also showed that matter and energy are also two aspects of the same thing. Furthermore, Kurt Gödel, a contemporary of Einstein's, discovered that the Universe is full of rotating matter, with time loops through every Point.

Intrinsic to Einstein's equations and models is relativity; the placement of objects relative to one another. It could be said that Einstein realized that space, time and the position of Being were all tied in together. Mendal Sachs maintained that relativity theory implies that, "the space and time coordinates are only the elements of a language that is used by an observer to describe his environment" (as shown in the visual examples in this chapter). So Einstein realized the importance of sentience, of Being, but he did not express the idea of Being as a measurable aspect of the Absolute in its own right.

Scaling and the impact of environment on objects

Galileo noted that if a physical object is scaled up or down in size, the resulting object does not have the same properties. While this does not apply to the scaling of geometric shapes and patterns, whose rules remain intact, as shown in Chapter 8 (fractals and patterns) and Chapter 13 (scale and matter),it does apply to physical forms of Being, which are essentially objects moulded out of compound pattern-sharing.

At a microscopic level there are numerous mandala-like organisms that are exceptionally fragile. Truly massive forms of Being, such as planets, revert to the sphere; their motions are guided by other geometry, such as the Spiral. On a smaller scale this process is evident when we take a rock and tumble it in water; eventually delicate edges and corners are eroded to form a smooth pebble.

Gravity is the ingredient that has an impact on scaling and other facets of Being. For example, tiny flies can walk up and down vertical walls, but if a fly was enlarged to the size of a dog the force of gravity would pull its mass downward and its spindly legs would collapse under its body weight. Intricate, fragile structures are only possible when they are tiny.

Scaling of any compound form of Being, or object, is determined by its environment. Forms of Being developed to survive in their given environment; in other words, their relative size in the grand scheme of things. The same holds true for humans. Our size is perfect for being able to make physical tools and use the elements, such as fire; something an ant could not do.

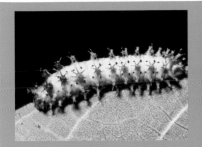  

Far left **The world comprises a flat plane for this caterpillar, which stays firmly on the surface of the leaf.**

Centre left **How would a praying mantis see its environment, since it can fly, yet still connects to surfaces on delicate legs?**

Above left **A gossamer web holds the weight of a dexterous spider.**

Far left and near left **Elephants, like large extinct dinosaurs, need solid legs to hold their weight. As a result they are not dextrous.**

Left **The typical human is perfect for manipulating the environment and the elements (earth, fire, water and air) for heat, tools, clothing and shelter. Having opposable thumbs helps, especially the ability to precision-grip, which is only found in the higher apes (and only in degrees significantly more restricted than in humans). Combining this with our intelligence and sentience we can utilize resources to adapt to our environment.**

The Prime Cross

According to Peter Plichta, in *God's Secret Formula: Deciphering the Riddle of the Universe and the Prime Number Code* (1997), Einstein's relativity equation connects the 3 most important physical dimensions we perceive with our senses: matter, energy and quantity and their "inverse" of space, time and numerical sequence.

Plichta maintains that these "inversions" are only accessible because of our ability to conceptualize and that the boundary between the finite observed reality and infinite realm of our imagination is the human Mind. He also postulates that if matter did not exist then nor would space and that where there is no movement there is no time. Therefore he argues that energy is the reverse of time.

Above **Square grids on the surface of the Earth bend, as can be seen when they are viewed from a longer perspective.**

"As above, so below"
According to some esoteric schools of thought, the legendary figure Hermes Trismegistus, the Thrice Great Hermes, was an initiate of the Egyptian Mystery School and the patron saint of alchemy. "As above so below", is a widely known quote from his alchemical text, *The Emerald Tablet*. Usually it is interpreted to mean, "that which is of the lesser world (the microcosm) reflects that of the greater world or Universe (the macrocosm)". Another way of putting it is that the structure of the largest is the same as the structure of the smallest, such as atoms compared to the solar system.

Scaling and perception

Scaling is all about size and also our perception of spatial measurement. On a very large scale the role of non-Euclidean geometry is more apparent since straight lines and forms, as seen in the formation of planets, become curved. Apparently straight parallel lines, as seen in a Square when we are on the surface of the Earth, bend when viewed from a higher perspective.

While compound structures may not scale up due to their mass and the effect of gravity, we do see patterns underlying form, scaling larger and smaller as they still utilize the geometric blueprint of the Matrix of Space-Time-Being.

Scaling down

Moving below the surface of things into the structure of matter takes us into a realm of wave-forms that crystallize as solids facilitated by geometry. Apparently solid forms comprise patterns and shapes that are intangible concentrations of vibrating waves made of energy. The appearance of its surface is due to the rapidly oscillating atoms.

String Theory

Mathematical Superstring Theory was developed in the 1980s to explain the properties of elementary particles in atoms and the forces between them in a way that combines Einstein's Relativity Theory, dealing with large mass objects in fairly large regions of Space-Time, with quantum theory and the subatomic scale. This theory is also known as The Big T.O.E. (Theory of Everything). So String Theory looks at minute scales of matter and how it behaves in Space-Time. It does not consider sentience. String Theory proposes that the fundamental building blocks in the Universe are not point-like objects or waves, but extremely small mobile string-like objects that exist in the Universe in *10* spatial "dimensions". Significantly, for reasons not yet understood by these theorists, only *3* spatial dimensions of length, width and height and a single-time dimension are discernable to us. But each of these spatial measures is made of *3* dimensions "enfolded" to appear as only *1* each. The tenth dimension is time. Even though

M-theory

M-theory was proposed in the 1990s to unify all the slightly different string theories. It asserts that strings are one-dimensional slices of a two-dimensional membrane vibrating in 11-dimensional space. M-theory also describes other Universes like ours, with 4 observable Space-Time dimensions, as well as alternate Universes with up to 10 flat space dimensions, and also cases where the position in some of the dimensions is not described by a real number, but by a completely different type of mathematical quantity.

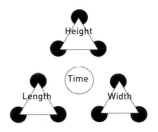

Above **String Theory.** Note the underlying Tetraktys formation and 3 x 3 = 9 Trinities.

String Theory comprises incredible, consistent mathematical equations, there is no practical evidence for it and it has not been confirmed by scientific research.

String Theory has been included in this book for two reasons. Firstly, at a quantum level it includes loops and strings; the same structures favoured by carbon. Open strings have two distinct end points and closed strings make a loop. The two types of string behave in slightly different ways. For example, in most string theories, one of the closed-string modes is the graviton and one of the open-string modes is the photon. Secondly, there is a striking resemblence between the String Theory model and the Tetraktys.

"Not that the propositions of geometry are only approximately true, but that they remain absolutely true in regard to that Euclidean space which has been so long regarded as being the physical space of our experience."

ARTHUR CAYLEY,
BRITISH MATHEMATICIAN
(1821–1895)

Calabi-Yau shapes

These images attempt to account for the additional dimensions required in String Theory. These unusual images rely on abstract modelling that many regard as unscientific.

Above **A 2-dimensional hypersurface of the fifth-degree Calabi-Yau three-fold.** Note the use of non-Euclidean geometry.

Above **In this Calabi-Yau, loops within loops are shown.**

Tetraktys and Pythagorean teachings

The Tetraktys described in Chapter 4 (see page 40) served as the basis of Pythagorean studies of natural science and philosophy. They used the 4 levels to represent the increasing density of the 4 physical Elements and 4 states of matter.

Descent from "fire to earth" as represented by 1, 2, 3 and 4 Points is a metaphor for the way the numbers unfold to 10 and become manifest in the many creations of Nature. The Tetraktys was also used to show the number of Points in Space of 0, 1, 2, and 3 dimensions that defined the Void, Point, line, plain surface and volume.

Pythagoreans saw this unfolding of the Universe from the Void represented in the same sequence of plant growth (seed, stem, leaf and fruit). Note the similarity of structure to the String Theory model (see page 120).

Point	Fire		Seed
Line	Air		Stem
Plane surface	Water		Leaf
Volume	Earth		Fruit

Fire   Air   Water   Earth

Seed   Stem   Leaf   Fruit

> **Tetrad**
> The principles of the archetype of 4 were called the Tetrad by the Greek
> mathematical philosophers and it is expressed as volume and 3D space in
> geometry and in Nature. There are 4 ways to examine solids – point, lines,
> area and volume, or as corners, edges, faces and from the Centre outward.

E8: "An Exceptionally Simple Theory of Everything"

Over 100 years ago an investigation of symmetry resulted in the fabulous E8
polytope (a geometric object existing in any number of dimensions). More
recently E8 has resulted in an innovative unified field theory proposed by physicist Garrett Lisi that attempts to describe all known fundamental interactions in
physics and hopes to stand as a possible "theory of everything". Unlike String
Theory, which requires extra dimensions to work, E8 attempts to describe all the
fields of Einstein's 4 dimensional model plus gravity as different parts of only one
field. In order to truly form a "theory of everything", Lisi's model must eventually
predict the exact number of fundamental particles, all of their properties, masses,
forces between them, the nature of Space-Time and the cosmological constant.

E8 encapsulates the symmetries of a geometric object with a "dimension" 248.
Note the doubling sequence of numbers 2, 4 and 8. The "dimension" described
in this symmetry is not spatial; it corresponds to mathematical degrees of freedom, where each "dimension" represents a different variable. What makes E8 so
exciting is that its foundation is geometric patterns within a Circle and symmetry.
It is embedded at the heart of many parts of physics and, so far, all the interactions predicted by the complex geometrical relationships inside E8 match with
observations in the real world. As we have seen, Nature likes geometry and symmetries. It is generally accepted that E8 research will have many implications,
most of which are not yet understood.

Lisi realized that he could place the various elementary particles and forces on
E8's 248 points. The coordinates of these points are the quantum numbers (the
charges) of elementary particles, which are conserved in interactions. Twenty
gaps remained, which he filled with notional particles, for example those that
some physicists predict to be associated with gravity. Lisi has noted how physicists have puzzled over why elementary particles appear to belong to families,
a trait which arises naturally from the geometry of E8. It may even be possible
to test Lisi's theory, which predicts new particles, by using the Large Hadron
Collider near Geneva, Switzerland to find them.

> *"Here is the world, sound as a nut, perfect, not the smallest piece of chaos
> left, never a stitch nor an end, not a mark of haste, or botching, or second
> thought; but the theory of the world is a thing of shreds and patches."*
>
> RALPH WALDO EMERSON, AMERICAN WRITER AND POET (1803–1882)

E8 polytope family graph
As a two-dimensional graph the E8 family is spectacular. Each coloured dot is the co-ordinate of a quantum number's point.

Above and right **Note that the graph's nodes fit inside a model that essentially comprises nested spheres. This is more apparent in E8 diagrams that have lines joining groups of related nodes, as in this E8 graph.**

Adding dimensions to the shapes

Scaling up and down in size relies on traditional three-dimensional space. We can comprehend this. What lies beyond the 3 dimensions of cube space? Can we conceive of reality being able to accommodate more than the three-dimensional spaces we can measure and perceive as volume? Any dimensions above 3 to locate a Point in space are abstract concepts, mathematically independent of the space the object exists in, as we know it. A whole area of mathematics is devoted to mapping geometric shapes called polytopes in higher dimensions beyond those of cubic space.

Let us consider the cube and process of cubing, since spatial measures are easiest for us to grasp. Cubed space is 2 x 2 x 2 = 8. The fourth step after cubed space is (2 x 2) x (2 x 2) = 4 x 4 = 16, 2 Squares that are themselves squared. If we continue the spatial sequence by moving the cube as a unit, a fourth-dimensional hypercube is made, also known as a toroidal hexagram and tesseract (see page 126). Its origination from, and its ties with, the Square Grid, are evident as its Squared central hole.

Continuing with the abstraction, just to see what happens to the number sequence, on the fifth step of multiplication, 2 to the power of 5 = 2 x 2 x 2 x 2 x 2 = 32 and 3 + 2 = 5. The Triangle of Being (3) with the interplay of Duals (2) together comprise the number 5 of the Spiral Life Force that animates and sustains life. This is reinforced by the first 5 numbers in the Fibonacci Series; 0 – the Void, 1 – the Point, 2 – Duals, 3 – Triangle of Being and 5 of the Spiral Life Force.

On the sixth step, 2 to the power of 6 = (2 x 2) x (2 x 2) x (2 x 2) = 8 x 8 = 64. The sum of 6 + 4 = 10 = 5 + 5 and 10 represents completion and the start of the next power of generation.

What about interaction of time, space and Being?

N-cubes, n-spheres and n-simplexes add measurements to their primary geometric shape in isolation of each other. They are abstractions of mathematical geometry that demonstrate that, in all possibility, geometry would structure reality beyond the confines of our perception in incredibly complex ways.

Note that these beautiful models essentially confine themselves to structure, form, or body, so they give us an insight into spatial complexity of form. They do not really give us further insights into the workings of higher dimensional time and Being.

Measurements, or dimensions, of time are temporal. Although time is often referred to as the fourth dimension in special relativity for this reason, time is NOT a spatial dimension. A temporal dimension is one way to measure physical change. It is perceived differently from the 3 traditional spatial dimensions in that there is only a single temporal dimension and that we cannot move freely back and forth in time.

The apparent perception of time flowing in one direction is an artefact of the laws of thermodynamics. Time is actually a network of interrelated events in the

> Dimensions are limitless, time is endless. Conditions are not constant, terms are not final. Thus the wise man looks into space and does not regard the small as too little not the great as too much, for he knows that there is no limit.
>
> Lao-Tzu, Chinese philosopher (6th century bc)

The Tesseract

- A generalization of the cube to dimensions greater than 3e is called a hypercube or n-cube.
- The Tesseract is also called a toroidal hexagram, an 8-cell, regular octachoron, 4-dimensional hypercube or 4-cube, 4-dimensional hypercube or 4-cube.
- Just as the surface of the cube consists of 6 square faces, the hypersurface of the tesseract consists of 8 cubical cells.
- The Tesseract consists of 4 cubes sharing 4 faces with 16 vertices.
- When viewed as a plane figure it is possible to see 4 Squares and a star with 8 points.

Top left **4-cube Tesseract (created by Robert Webb's Great Stella software).**

Left **The 4-cube is drawn by translating the cube.**

Sequence of plane graphs of the n-cube:

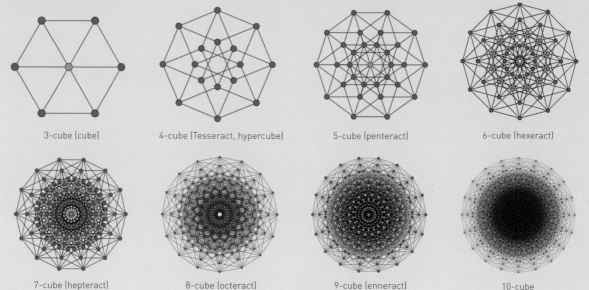

3-cube (cube) 4-cube (Tesseract, hypercube) 5-cube (penteract) 6-cube (hexeract)

7-cube (hepteract) 8-cube (octeract) 9-cube (enneract) 10-cube

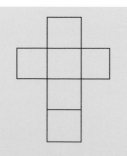

Three-dimensional Latin Cross

Just as the cube can be unfolded into 6 Squares, the Tesseract can be unfolded into 8 cubes to form a three-dimensional Latin Cross.

Unfolded cube into 6 Squares   Unfolded Tesseract into 8 cubes

n-dimensional simplex

The series of mappings for the Triangle are known as n-simplexes.

- 0-simplex is a single point
- 1-simplex is a line
- 2-simplex is a Triangle

Left **5-cell is a 4-dimensional object bounded by 5 tetrahedral cells. It is also known as the pentachoron, pentatope or hyperpyramid.**

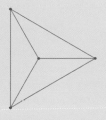

3-simplex (tetrahedron)

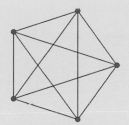

4-simplex (pentachoron)

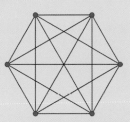

5-simplex (hexateron)

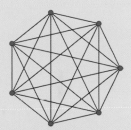

6-simplex (heptapeton)

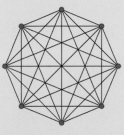

7-simplex (octaexon)

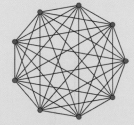

8-simplex (enneazetton)

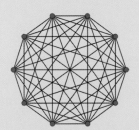

9-simplex (decayotton)

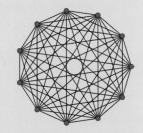

10-simplex (hendecaxennon)

Hyperspheres

1-sphere is a Circle in the plane
2-sphere is an ordinary sphere in 3-dimensional space
3-spheres is a hypersphere
Spheres of dimension n > 2 are sometimes called hyperspheres, or glomes

"For an object under the eye will appear very different from the same object placed above it; in an enclosed space, very different from the same in an open space."

MARCUS VITRUVIUS POLLIO,
ROMAN WRITER, ARCHITECT AND ENGINEER (c80/70–c15 BC)

Pentagon, octahedron, dodecahedron and icosahedron in 4 dimensions

These are some examples of abstract 4-dimensional structures for the remaining Platonic solids.

Above **4-dimensional dodecahedron** is the dodecaplex. Much as a dodecahedron can be built up as a model with 12 pentagons, 3 around each vertex, the dodecaplex can be built up from 120 dodecahedrons, with 3 around each edge.

Above **4-dimensional icosahedron** is the hexacosichoron. It has 5 tetrahedra meeting at every edge, just as the icosahedron has 5 Triangles meeting at every vertex. It is also called a tetraplex.

Above **4-dimensional octahedron** is the octaplex or octacube. It has 24 octahedral cells, with 6 meeting at each vertex and 3 at each edge.

Moment with no definite "direction" attached to them, like a map of dynamic potential that cannot be readily drawn as abstract visual models.

Scale of sentience, of Being

The 4 dimensions of space and time are needed to determine a unique vector for a specific event in the Matrix of Space-Time-Being. Each vector is a vantage Point in space, in a unique Moment in Time, where a Being in existence can "see" its Self relative to everything else. This act of "seeing" can be symbolized as an Eye of the Mind that looks out and perceives itself to be part of unfolding events. As discovered by quantum physicists, each event occurring in a given Moment-Point-Eye is not separate from all the other events in Space-Time-Being.

Sentience is the measure of Being, not physical size. Sentient microscopic animals would perceive their world in a totally different way to a sentient Being the size of a planet. A minute creature, like an ant, sees details that are lost on a vast scale. Very small creatures may see their environment in only two dimensions, as flat space, because they are not large enough to perceive the whole solid. Similarly, different life forms have very different perceptions of the passage of time.

As humans we have a unique vantage point with regards to physical scale. We can manipulate our environment and make tools to enhance our physical vision up and down the scales of size by using telescopes and microscopes. This increases our awareness of physical reality and scale. As seen we can also conceptualize abstract geometry.

The basis of Being on different Planes have been known intuitively for many

Spirals and cycles of the Universe

Atoms are highly compressed Spiral vortices of energy. Nuclear explosions are the phenomenal release of energy experienced as these Spirals of energy unwind with tremendous force. Imagine the Spiral power of the compressed energy in the Point that was released in the Big Bang. If the Universe collapses the Pulsating (or Inflating) Universe Theory suggests that it will explode again, producing a new expanding Universe, which will in turn collapse ad infinitum.

The spiralling aspect of the Universe appears in mythology as Serpentius, coiled around a staff representing the column holding up the Universe and the central axis around which life rotates (see page 52). The Spiral of manifestation around the central axis is also representative of the "breathing" cosmos. With exhalation, spirit contracts, creates and winds into matter, depicted as creation through the "breath of God". On inhalation, matter expands and evolves, unwinding into spirit. Theoretically, our Universe oscillates through expansion and contraction, like the interplay of Yin and Yang in the T'ai Chi symbol.

Eternal Return

An idea known as the Eternal Return describes how the Universe repeats as one giant Circle in Toroidal, doughnut-shaped Space-Time. Leading cosmologist J. A. Wheeler describes the structure of Space-Time in terms of the "vortex ring". In Wheeler's model the Universe Spirals around in a continuous cycle, yet there is no last time or next time because nothing is moving. Space-Time is structured on this model and it exists in a timeless state in which only our conscious experience of each Moment creates the illusion of the movement of time as a linear sequence of Moments.

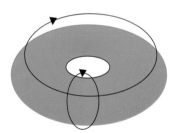

Above **Toroid vortex ring. Space stretches and time shrinks. Time is represented by the vertical ring and Space by the horizontal ring.**

generations. The higher Planes of finer vibration are considered to be sentient, according to many beliefs. Physical measurement of something as intangible as higher Planes is not feasible. Instead we use visual and verbal metaphors, such as the Eye and angels.

From the perspective of life as a human Being it is difficult to visualize and communicate abstract higher dimensions to each other. Instinctively we know these wider measurements are out there, seemingly beyond our confines, extending our reality into a much bigger reality of which we are but a small slice. But our slice is special, artfully constructed out of precise geometry just so that we can experience the joy of sentience in a physical form, in 3 spatial dimensions and time.

MEDIUM OF SOUND

Music is the wine which inspires one to new generative processes, and I am Bacchus who presses out this glorious wine for mankind and makes them spiritually drunken.

LUDWIG VAN BEETHOVEN, GERMAN COMPOSER (1770–1827)

Part III – Medium of Sound – explores the fundamental role sound plays in facilitating the creation of reality. You will learn about sound waves, the geometry carried within sound and why certain sounds are regarded as sacred. The art of sound is music and you will discover why music can have such a profound effect.

15 Sound as a vehicle for geometry

In the beginning was God, with whom was the Word; and the Word was truly the supreme God.

THE BIBLE, JOHN 1:1

SOUND AND HEARING PRECEDE LIGHT AND SEEING. In accordance with this, most of the world's religions and myths indicate that creation began with, and was enabled by, sound: "God commanded the light into being through the medium of sound". How is sound the enabler of creation? Sound is crucial because it is activated by an event changing the status quo. Stillness is set in motion and now the geometric code can be carried on the undulating waves of sound that move through anything manifest.

Big Bang echoes

Reported in May 2001, an international team of scientists claimed they had detected the sound of the creation of the Universe. They picked up the echoes of the Big Bang by surveying the sky over Antarctica. These echoes are remnants of the huge acoustic waves that surged through the white-hot plasma gases generated in the explosive birth of the new Universe. They suggested that these waves "shaped" the Universe by concentrating matter in some areas and removing it from others. They also believed that the results are confirmation of the Expanding Universe Theory (see page 50). If the sound had not been there the Universe would have been filled with a uniform gas.

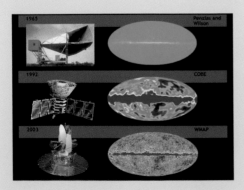

Top left **Penzias and Wilson's simulation of the sky viewed by their microwave receiver (1965). They discovered the green remnant afterglow from the Big Bang and were awarded the Nobel Prize for their discovery (NASA).**

Centre left **The COBE Spacecraft and the painting of COBE's view of the early universe (1992). COBE first discovered the patterns in the afterglow (NASA).**

Below left **The WMAP Spacecraft and a computer rendering simulating WMAP's view of early universe (2001). WMAP bring the patterns into much better focus to unveil a wealth of information about the history and fate of the Universe (NASA).**

Audible and silent sound

Without a medium, sound is not possible and this is why it cannot travel through a vacuum. Sound is produced as a result of changes in pressure, making waves or oscillations that move through a physical medium of liquid, solid, gas and plasma.

An ability to detect these pressure changes in the air is useful for navigation and for perceiving danger. Human ears can sense some sound waves, but certainly not the full range that is actually going on all around us and through us – continually. Anything that moves sets off waves and it is an important characteristic of Nature's creations that they are continually in motion, producing their own unique audible sounds as they interact with their environment – wind, water, fire, animals, plants and even our nostrils and stomachs. In addition, every form of physical Being – trees, furniture, animals, clothing, people and planets – have sounds that are not audible. Those sounds are produced by their molecules and even between the molecules, because all matter vibrates.

The concept of "silent" sound is well understood by Native Americans. They call animals to the hunt and find plants using sound that is not formed by vocal chords but by their own energy fields – by creating an image of the form in their Mind and calling them, using what is termed their Higher Self.

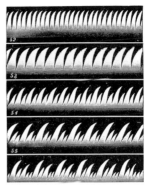

Above **Tonal sound variations' effect on a candle flame (*Popular Science Monthly*, 1878).**

Sound waves in Nature

Waves produced by sound are mirrored by the undulations in water due to an event setting off a rippling "wave", which moves the water particles up and down. In turn, distinctive creases are left in the wet sand and mud because of the oscillating motion of wavelets across the beach.

The same effect is seen in desert sand dunes created by the wind – sound arises as the sand is blown by the wind. All these natural phenomena demonstrate the interdependent cause and effect of "waves" of energy and continual motion in Nature.

Above **Illustration of the undulating motion of compressed, rarified air particles of sound waves.**

Above left to right **Ripples in water, mud, sand and in the clouds due to air currents (NASA).**

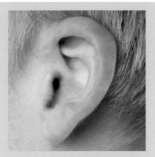

Above **The outer ear gathers in sound waves.**

Right and centre **Diagrams of the middle and inner ear.**

The structure of ears

Since the Life Force is behind motion and hence sound, it is not surprising that the Spiral has a critical role within the anatomy of the human ear for picking up and hearing sound.

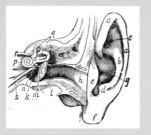

Cross-section of the ear's cochlea

Below **Ears come in all shapes and sizes in the animal world. Ears are an essential sensory tool for sentient animals.**

Ears

Like humans, other vertebrates hear primarily using the auditory system, where vibrations are detected by the ear and transduced into nerve impulses that are then perceived by the brain. Other mechanisms apart from ears may be used. For example, snakes sense sound through their bellies. Not all sounds are audible to all animals and the ranges that are detected vary. Dolphins have a small ear opening on each side of the head, but it is believed hearing under water is also, if not exclusively, done with their lower jaw, which conducts sound to the middle ear via a fat-filled cavity in the lower jaw bone.

Broadcasting sound

As an extension of our faculty to make sounds for communication, pleasure and spiritual purposes, we humans have invented a wide range of tools and technology to generate, record, transmit and broadcast sound across huge distances, almost instantaneously, to huge audiences.

The Hum

A low-pitched sound called the Hum can be heard in numerous places across the globe, especially in the USA, UK and in Northern Europe. The Hum is sometimes prefixed with the name of a place where the phenomenon has been particularly publicized (for example, the "Bristol Hum", the "Taos Hum" or the "Bondi Hum"). Said to be similar to a distant diesel engine, you need a very quiet environment to hear it. Unlike the sound vibrations caused by an event, such as a volcano, which fade, the Hum is continual. These sorts of vibrations are called "background-free oscillations" because they keep on sounding even after their original source has disappeared. The Taos Hum is said to be the natural tone of Planet Earth. Another theory is that it is the atmosphere pounding the Earth like a drum or gong.

In 1998 the Japanese geophysicists Naoki Suda and Kuzunari Nawa claimed to have isolated the Hum in seismic data. It comprised 50 notes crammed into less than two octaves, lying about six octaves below middle C. Pitches of these notes range between 2 and 7m Hz. Each individual note is pleasant, but played together can have negative side effects. Overexposure can cause dizziness, shortness of breath, headaches, anxiety and sleeplessness. Symptoms like these are commonly found among people living in various cities in New Mexico, USA.

Deliberately making sound

The role of sound and the ability of Beings to make it is highly significant. Many species, such as birds, cats and dogs, have specifically developed organs to produce sound, just as humans have a larynx (see also page 142). Sounds unique to the species allow individuals to communicate, to search out and woo a mate. As a vehicle for communication and for sharing ideas, sound lifts a species up the evolutionary scale. Sound as a creative medium works on physical and spiritual levels, which is why repeating sacred words with intent invokes powerful forces and why music, singing and chanting are significant in rituals. We sense how sound affects mood and body on a more subtle level; lifting spirits when we are unhappy, energizing us through rhythm and establishing an atmosphere.

The human larynx

Above **Illustration of Euler's method and apparatus for working out his "wave equation".**

Leonard Euler's wave equation

In 1748, the Swiss mathematician, Leonard Euler, worked out the "wave equation" for a string that involves rates of change relative to time and to space. Euler demonstrated how two waves form, complementing each other, with one travelling one way and the other in the opposite direction. The physical laws for light, sound, electricity, magnetism, the elastic bending of materials, the flow of fluids and chemical reactions are all equations for various rates of change.

The same equation for waves features everywhere, showing the hidden unity of all things. For example, James Clerk Maxwell, a British physicist, showed that electric and magnetic fields affect each other in such a way as to allow waves to travel through space. His equations matched those that describe light, though they were more general, covering electromagnetic phenomena other than light.

Right **Diagram showing different pitches of sound waves.**

Far right **Wave forms in water.**

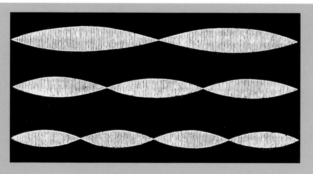

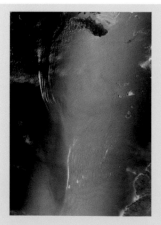

Atmospheric rippling

The wave pattern shown here is of atmospheric gravity waves on the surface of the ocean. They form when buoyancy pushes air up and gravity pulls it back down.

On its descent into the trough of the wave the air touches the surface of the ocean, pushing the water into ripples, like those seen in sand and mud. The brighter regions show the crests of the atmospheric waves. Beneath the crests, the water is calm.

Left **Satellite view of Shark Bay, western coast of Australia.**

Creative 3 in music

Using a vibrating string or air column of different lengths it is possible to put together the 7 note musical octave out of just 3 combinations of 3 primary musical notes (the tone, 4th and 5th) and their 4 secondary waves (see also Chapter 17).

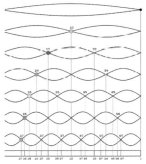

Above **Standing waves** are seen in physical media, such as strings and columns of air. Any waves travelling along the medium will reflect back when they reach the end. This is most noticeable in musical instruments where a standing wave is created, allowing harmonics to be identified.

Left **Opening mouths** literally and symbolically exhibit the deliberate use of the "voice" throughout the animal kingdom on the surface of the Earth. Dolphin sonar and whale song under water are also forms of communication using sound wave.

Above **Aurora Borealis, Alaska.**

Magnetic rope section

A Birkeland current is a magnetic field aligned current in the Earth's magnetosphere (which is formed when a stream of charged particles, such as the solar wind, interacts with, and is deflected by, the magnetic field of a planet or a similar body). The current flows toward the Earth on the dawn side and in the other direction on the dusk side of the magnetosphere. The current flows parallel to the Earth's magnetic field. When it reaches the upper atmosphere of the Earth it produces the Aurora Borealis and Aurora Australis.

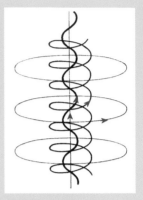

Above **The self-constricting magnetic field lines and current paths in a Birkeland current, or magnetic rope.**

Whirling

Twirling skirts of ritual dances, such as the Dervish and in other cultures, mirror the magnetic whirls around a sending wire and around the Sun due to its magnetic field.

Far right **The undulating surface, like a skirt, within the Solar System where the polarity of the Sun's magnetic field changes from north to south is called the helioscopic current sheet (NASA).**

Right **Magnetic whirls around a sending wire.**

Far right **Ritual spiritual dances often involve spinning so that skirts flair out.**

Right **The skirt of the Whirling Dervish imitates the pattern of the helioscopic current sheet.**

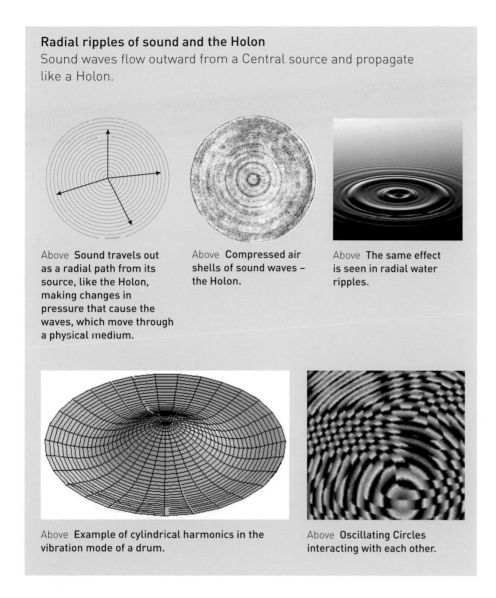

Radial ripples of sound and the Holon
Sound waves flow outward from a Central source and propagate like a Holon.

Above **Sound travels out as a radial path from its source, like the Holon, making changes in pressure that cause the waves, which move through a physical medium.**

Above **Compressed air shells of sound waves – the Holon.**

Above **The same effect is seen in radial water ripples.**

Above **Example of cylindrical harmonics in the vibration mode of a drum.**

Above **Oscillating Circles interacting with each other.**

Cymatics and geometry hidden in sound

What does sound have to do with geometry? Sound has a vital role to play in the formation of reality and this is due to a hidden feature of the sound waves, a feature only recently revealed to the naked eye. Forms that appear solid are actually created by their underlying vibration, as is revealed by the artworks of cymatics.

The term cymatics derives from the Greek *kuma* meaning "billow" or "wave" and attempts to describe the effects that sound and vibration have on matter. Cymatics reveals sound vibrations in the form of images and it falls within the discipline known as "acoustics". Cymatics, broadly speaking, looks at how vibrations generate and influence patterns, shapes and moving processes. Not only does cymatics demonstrate that sound affects physical matter, but, more importantly, that sound creates geometric shapes by virtue of its action.

139

Seeing geometry in sound

German physicist and musician Ernst Chladni led the way to understanding the link between energy and geometry. In the mid-18th century Chladni made sound waves "visible" by drawing a violin bow across flat plates covered in sand, to produce beautiful patterns known as "Chladni figures". Two centuries later, in the late 1960s, a Swiss doctor called Hans Jenny took various materials (for example iron filings, spores and water) and placed them on vibrating metal plates. The resulting patterns varied. While some were perfectly ordered and static, other patterns continually changed and moved. The range of viewing cymatic patterns has been extended recently by the use of a light microscope.

Dr Jenny's results showed that characteristic patterns could be produced for different materials and that by changing the frequency or amplitude of sound the patterns changed. Sound therefore has a measurable effect on matter. By increasing the frequency (more waves per second) the complexity of the pattern formed increased. Greater amplitude (making waves bigger) resulted in equivalent motions that were more rapid, and even quite violent. Jenny described material being thrown off the plate in volcano-like eruptions. Under certain circumstances Jenny could make the patterns change without altering amplitude or frequency. An interesting phenomenon occurred when liquids were used as the medium on the plates. If the plates were vibrating the liquid stayed on the plate when it was tilted and new shapes continued to be built. If the oscillator was turned off the liquid would then run off the plate unless the machine was quickly turned back on again. So, sound vibrations also caused anti-gravitational effects.

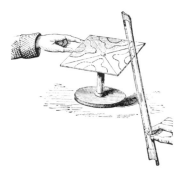

Above **Chladni's original method for producing Chladni figures.**

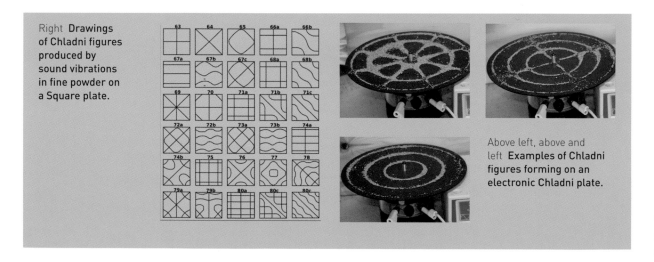

Right **Drawings of Chladni figures produced by sound vibrations in fine powder on a Square plate.**

Above left, above and left **Examples of Chladni figures forming on an electronic Chladni plate.**

Cymatics and E8

Significantly, the counterpart to Jenny's research in cymatics in the field of subatomic particles is the E8 model. There are similarities between the patterns of cymatics and quantum particles. In both instances, what appears solid is in fact a wave form and created and organized by vibration.

Sonic spheres

John Stuart Reid (1948-), an acoustics engineer, currently researches cymatic phenomenon and he proposes that a source instigates the ripple as an event in time and that audible sounds are spherical sonic bubbles, in form, not longitudinal waves, as is commonly believed. His CymaScope (see below) shows us a two-dimensional slice through the sonic bubble, revealing the interior structures of a given sound. Sound propagates spherically in air due to diffraction, the reactive result of atomic collisions. Reciprocal effects in air occur in the jostling of molecules initiated by a sound event, causing components of the sonic energy to move in all directions almost simultaneously. The distribution of energy within the sonic bubble is always concentrated on an axis with the direction of original transmission from the sound source.

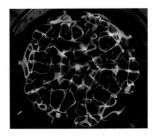

Above **Symmetry in a vibrating fluid.**

The original CymaScope; the drum

At least 1,000 years ago African tribes used the vibrating skin of drums sprinkled with small grains as a divination method for foretelling future events. As the drum is one of oldest-known musical instruments, the patterning effects on sand of a vibrating drum head have probably been known for millennia by other cultures as well.

Above **Geometric cymatic wave patterns caused by sound vibrations in water.**

Invention of the CymaScope

John Stuart Reid conducted cymatics research in the King's Chamber of the Great Pyramid of Egypt in 1997 by stretching a PVC membrane over the sarcophagus. As many of his CymaGlyph images strongly resemble Ancient Egyptian hieroglyphs, Reid postulated that the inherent resonances of granite, when radiated as a complex sound field during the crafting of the stone, might have influenced the development of hieroglyphic writing.

Reid subsequently began experimenting with instrumentation that would enable an accurate visual equivalent of sound to be created from any audible sound, resulting in the invention of the CymaScope. Working with Erik Larson, he designed and built the instrument to Pythagorean proportions. Reid recognized that the wave model of sound is incomplete because it does not characterize the spherical space-form of audible sound. CymaGlyph images, created on the CymaScope, are considered to be analogues of sound and music, since the geometries they contain correlate mathematically to the musical pitches that caused the pattern to form on the instrument's membrane.

16 Sacred sounds

The sound is fading away
It is of five sounds
Freedom
The sound is fading away
It is of five sounds

CHIPPEWA SONG

DR HANS JENNY WAS UTTERLY CONVINCED that our biological evolution occurs as a direct result of vibrations (see also page 141). He believed that each type of cell has its own frequency and that a group of cells with the same frequency in turn creates a new frequency that is in harmony with the original one. This belief was based on the striking similarity between the shapes and patterns found in the pattern-sharing evident in Nature and those he generated. Jenny also maintained that close study of the human ear and the larynx would provide a better understanding of the importance of vibrations.

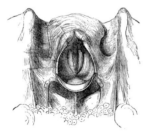

Human vocal communication

The versatile human larynx and our advanced neural abilities combine to enable us to produce sounds such as speech, humming, singing and shouting. This unique ability has been used by humans to access the spiritual Planes of Being through praying, toning, humming, in meditation, chanting and other vocal techniques.

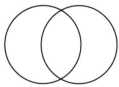

Top and above **The human larynx is very similar in shape to the female yoni and the vesica piscis, both symbols of female generative power.**

Divine breath manifesting archetypes

The symbolism that originated on cave walls (see page 11) gradually developed into writing when large urban populations settled in China, Mesopotamia and Egypt, replacing nomadic cultures. As representations of the sounds we make, rather than as abstract symbols of reality, the earliest records of writing followed our ability to make sounds using our larynx by millions of years. In these early writing systems Asian Indians focused on the sound and structure of language and the Chinese on patterns with ideas attached to them, known as ideograms.

To copy the alphabet used to be, and in some cultures still is, an act of worship. Other cultures also regard the sound of letters and song as sacred. For example, Australian aborigines regard their songs as particularly important – they believe

"To think is to talk to oneself and each of us talks with himself thanks to our having had to talk with one another."

MIGUEL DE UNAMUNO, BASQUE NOVELIST, POET, PLAYWRIGHT AND PHILOSOPHER (1864–1936)

them to have magical powers indivisibly linked with the natural world. Letters of sacred texts are, like geometric models, objects of contemplation that unlock the power of creation by virtue of their ability to work together to become something else. Sounding out these holy letters through a conversation, by reciting poetry or chanting mantras, singing songs, shouting a curse or crooning softly, we give sound power and it "lives".

In Islam the Divine Exhalation is the "manifestation of the Creative" and is comparable with the creative power of the Hindu goddess Sakti. Divine Archetypes, or names are considered to be created through God's divine breath via the 28 letters of the Arabic alphabet. These letters correspond to the shapes made by the waxing and waning of the Moon. They are also phonic and as such their form, sound and inner meaning as divine names are very closely related.

Hindu tantric philosophers devote a large portion of their literature to the explanation of sacred sound, symbol and worship. They declare that the Supreme (*para*) stirs forth multiple forces of existence through the Word (*Sabda Brahman*). The process of creation from the subtle ether to the form happens as cosmic vibrations (*nada*). The degree of vibration varies in concentration and wavelength, creating that which we perceive as light, volume, mass and structure.

As Alexandra David-Neel relates in *Tibetan Journey*, a Tibetan priest she met called himself a "master of sound". His view was that all things are an assembly of atoms that "dance" and because of this they produce sounds. Naturally any change in rhythm in their "dance" changes the sound. Each atom even has its own "song" working harmony with other atoms to make both dense and subtle forms of Being. And, according to field theory in physics, each atom produces rhythmic patterns of energy as evident in their waves of varying frequencies of vibration.

Top and above **Opening our mouths into a Circle reveals the inner hole (like the Central Point) from which sound comes out to create noise. By making noise we set off "creative" vibrations that effect any physical Being they ripple through.**

Geomatria, the integration of language and number

Language, and hence, sound, were integrated into numbers in Ancient Greek, Egyptian, Hebrew, Arabic, Coptic, Syriac and Sanskrit. Letters had number values attributed to them to reveal their sacred symbolic significance. For example, the sum of the numbers for the Greek name for the Egyptian River Nile totals 365 in reference to the 365-day cycle of the Nile's rising, flooding, depositing of rich mud and receding.

ΚΟΣΜΟΣ (cosmos) is Greek for universal order. It has a sum value of 600 to highlight the importance of 6 and its multiples.

Above **Zen artist Wang Xizhi created a manuscript of a Taoist passage in AD 356. This image is of part of a Song Dynasty stone rubbing of that manuscript.**

Examples of sacred alphabets

Arabic alphabet

Hebrew alphabet

Sanskrit alphabet

Living calligraphy

Calligraphy images are viewed as pictures of the Mind and our degree of enlightenment is displayed, expressed in the flow of the ink. Someone once tried to rub away some characters written by Wang Xizhi (a Zen Buddhist master calligrapher who lived about 200 years before Bodhidharma came from the West) and found that the characters he had written had penetrated the wood on which they were inscribed and were impossible to erase. Studies by Terayama Katsujo in Japan have shown that Zen calligraphy is "alive". He collected originals, fakes, copies and modern calligraphy work. Under magnification using an electron microscope the ink particles of the original works were, in his view, "numinous", whereas forgeries and copies emitted very weak energy.

Om/AUM as the original sound source

Hindus believe that Om (also written as AUM) is the most ancient sound symbol for the "original source". Om acts as a non-specific name for the ultimate power behind life and is believed to be the original sound of creation. Om/AUM is believed by Hindus and Buddhists to be the primordial "Mother of all sounds"; the first breath of creation. It corresponds to the Egyptian "Amen". Originally Om/AUM was written in Sanskrit at a time when letters were viewed as symbols of the divine and the act of speaking sacred letters invoked their latent power.

OM/AUM REPRESENTS THE TRINITY OF DIVINITY				
A	Brahma	Creator	Past	Conscious or waking
U	Vishnu	Preserver	Present	Dreaming: opening and closing the mouth
M	Siva	Destroyer	Future	Dreamless sleep: sound made with lips closed

The Om/AUM symbol

Together, as A-U-M, the 3 letters become an eternal syllable that transcends time. The entire word, encompassing the crescent and the dot, stands for a fourth state of consciousness called *samaditi*; a state lying beyond the limitations of shape and form that is described as pure white light. A very interesting similarity worth noting is that according to the Navajo of North America our world is the fourth white world. *4* sequential sand paintings are ritually drawn as an accompaniment to chanting parts of the Navajo Emergence Myth. These sand paintings are in essence a re-enactment of creation.

Siva Nataraj's dance
Through symbols and dance gestures, Siva teaches that he is Creator, Preserver and Destroyer. Through him, the soul of mankind can be transported from the bondage of illusion and ignorance to salvation and eternal serenity. Encircling Siva is a flaming body halo (*prabhamandala*) that symbolizes the boundaries of the Universe. Siva Nataraj (Lord of the Dance) has the following features:

Siva Nataraj

- In his upper right hand is the *dammaru*, the hand drum from which issues the primordial vibrating sound of creation.
- With his lower right hand he makes the gesture of *abhaya*, removing fear, protecting and preserving.
- In his upper left hand Siva holds *agni*, the consuming fire of dynamic destruction.
- With his right foot he tramples a dwarf-like figure (a*pasmara purusha*), personification of illusion, who leads mankind astray.
- In his dance of ecstasy Siva raises his left leg, and, in a gesture known as the *gaja hasta*, points to his lifted leg to provide sanctuary for the troubled soul.

> I will tell you the word that all the Vedas glorify, all self-sacrifice expresses, all sacred studies and holy life seek. That word is Om.
>
> KATHA UPANISHAD (C5TH CENTURY BC)

AUM and the number 5

AUM can be broken down into 5 parts that directly correlate with the 5 Elements (earth, water, air, fire and ether). Siva's eternal energy is made evident in 5 activities and his mantra has 5 letters. Each of the 5 seed mantras, single syllables that are symbols of an attribute or quality of the divine, also relate to the 5 parts of AUM.

The Saradatilaka Tantra describes a World Tree, *lipi-taru* (also known as the Cosmic Tree), made of a mesh of Sanskrit letters that are perceived as resonating vibrations of Universal energy. The Universe as composed of the 5 Elements is represented by sound combinations on various parts of the World Tree. In this way the physical world is represented by Sanskrit sound vibration-based equations.

5 and the underlying structure of form, as well as the replication of patterns through the Spiral are evident again, but now in the context of the medium that carries them: sound.

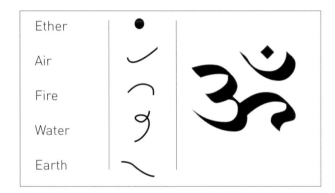

The Duals of sound and silence

Looking at AUM written as "Om", O is the Void, as symbolized by an empty Circle, and M can be regarded symbolically as looking like the wave vibrations that underlie ethereal and physical forms of Being. Every form of Being comes from this nothingness of the original Circular vessel and every Self resides in the Point at the Centre of the Circle, surrounded by the emptiness of the Void. Most physical forms of Being comprise empty space; the vast tracts of nothingness between atoms, like the empty space between the stars. Perhaps we pay too much attention to the forms and noise rather than the true essence of things: emptiness and quiet.

Like everything metaphysical the harmony
between thought and reality is to be found
in the grammar of the language.

LUDWIG WITTGENSTEIN,
AUSTRIAN PHILOSOPHER (1889–1951)

Outer space and silence are external manifestations of inner space and inner silence. Silence and emptiness are not objects of knowledge like things we can touch and hear; their essence is a state of knowing. Sound cannot exist without silence and only when sound exists does silence come into being. So when form exists space naturally occurs as well.

OM – Circle and vibrating wave

AUM when written out as angular forms is like vibrating waves in a Void:

HU

First there is total silence; the Circle of the Void. Then preceding the sound AUM is the breath, HU, made without using the vocal chords. It is the open mouth, 0, with the passage of air as the dot. HU is an ancient name for "the originator of life" according to Sufi tradition, and the first sound of the spiritual world rather than the material, realized world that AUM precipitates. Once we tone AUM we use our larynx to make the sound with the geometric code in it.

Om CymaGlyph created on the CymaScope instrument

"The vowel 'O' creates a series of simple concentric Circles, while the periphery is created by the 'm' sound, which is different for each person, depending on voice harmonics." (J.S. Reid)

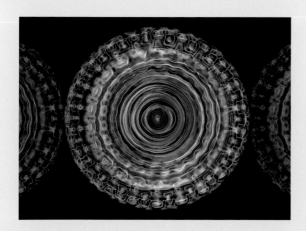

J.S. Reid Cymatic image (contributed by John Stuart Reid/CymaScope.com)

147

17 Music, rhythm and harmony

Above **A crowd clapping in unison heightens emotions and a sense of participation in the event (in time). Sound and event share the symbol of a Circle with a central Point.**

> When one looks at the cosmos, the movement of the stars and planets, the laws of vibration and rhythm, all perfect and unchanging, show that the cosmic system is working by the laws of music, the law of harmony...and it is these that Pythagoras called the 'music of the spheres'.
>
> HAZRAT INAYAT KHAN, SUFI TEACHER,
> (1882–1927)

> Silence is more musical than any song.
>
> CHRISTINA ROSSETTI,
> ENGLISH POET
> (1830–1894)

THERE HAS NEVER BEEN A CULTURE WITHOUT MUSIC. Some of the earliest-known instruments are over 20,000 years old, found in ancient settlements in north-western Europe and include decorated flutes made from mammoth bones and simple percussion instruments such as castanets. It is easy to see how drawing and painting may have originated from the need to communicate by recording stories to pass down knowledge and experience, but what utilitarian benefit can music possibly have had in our earlier evolution?

Possibly the first man-made music would have been a spontaneous act with a social role, such as clapping and drumming, or to heighten emotions before a cry to hunt or to go into battle. From humble beginnings music has become far more structured and formal, encompassing a huge range of sound levels, frequencies, rhythms and tempos, in many different combinations.

Left to right **The Franco-Prussian War (1870-1871); A Sioux war dance (1888); 13th-century Saracen army. Drums are universally used as a call to arms.**

Examples of primal drumming to the beat of Nature across cultures and ages. Drums have always played a central role in human ceremonies and gatherings of all kinds.

Expressing the inexpressible

Music is one of the few things that humans enjoy for pleasure and can appreciate without being an expert. From the moment of birth humans respond to music (even before birth, in the womb). Even as young as five months of age human infants are aware of tiny shifts in music and by eight months they can remember simple melodies. Without consciously having to learn the rules of music we respond instinctively, moving with the beat and yielding to its power to enhance emotions, to calm and to excite, to sadden or to make joyous. Perhaps our innate musical ability originated in the desire to make contact with the spiritual realm and the gods. At another level music is very personal, reaching right into our hearts. It vibrates body and soul. But there must be more than this; our inherent musical appreciation and the variations in our ability to produce and respond to music are too significant to just be by-products of our evolutionary process.

The rhythms of life

All aspects of musical form are "shaped" by the relationship between time and the rhythms of life. Rhythm is used to describe the time structures and tempo determines the speed of the beat. Like the rhythms in Nature, the beating of the heart and the seasons' cycles, musical rhythm organizes notes into regularly occurring patterns. These patterns regulate the motion of the music. Not surprisingly, we build our bodies' rhythms into our musical creations in various ways. The most basic is the beat, which is like the ticking of a clock and the beating of our own pulse. It can be overt, as in popular music, or hidden, as in much classical music. Melodies are divided into musical phrases that tend to produce intervals of time similar to the human breathing cycle. Such timing is also central to any human activity requiring co-ordination of the brain with eyes and limbs, for example, running, throwing a ball and, obviously, playing a musical instrument.

> Music in the soul can be heard by the Universe.
>
> LAO-TZU,
> CHINESE PHILOSOPHER
> (6TH CENTURY BC)

Above **Pythagoras studying the harmonics of instruments, shown in a woodcut, *Theorica musicae*, by Franchino Gaffurio (1492).**

Life in harmonic progression

Using the growth patterns of sunflower seeds, architect and writer Gyorgy Döczi shows that the diagonal lengths of seed groups form in the 5:8 harmonic relationship. By counting the seeds in neighbouring groups, a series known as a "harmonic progression" can be found.

Celestial harmony

Proportions play a fundamental role in the balance and harmony of geometry and proportions are fundamental to the generation of the Spiral Life Force that courses through life (see pages 46–53). Music demonstrates the same concept through harmony in sound. Harmony comes from the Greek word, *harmos*, meaning; "fitting, orderly, pleasant, a joining together of different things to make a group which works well together, such that they are fused and no longer separate". This applies to body proportions, art, design and Nature's creations. We are drawn to harmony (see page 65)

Pythagoras was fascinated by the study of music because it embodied numerical relationships. He discovered that tuned strings on a musical instrument sounded pleasant, or in harmony, when their lengths were in unison; related to each as equals, but with pitches an octave apart. The strings' lengths are then divisible by 2, or in 1:2, 2:3, 3:4 (1, 2, 3, 4) proportions of length; note the key role of numbers *1, 2* and *3*. These proportions also appear in the strongest overtones, known as "partials" or "harmonics". Harmonics are found blending in every single musical sound, like additional invisible strings being sounded at the same time. These additional harmonics turn mere noise into the art form "music".

Above **James Gillray's *Harmony Before Matrimony* (1805).**

Harmony and the Duals

When 2 opposites are in perfect harmony they can be represented by the T'ai Chi symbol (see page 26). If they are not in harmony we see discord. In this caricature (see left) James Gillray depicts a musical courtship incorporating symbols of harmony and discord. The man and woman, the Duals, sing a duet from "Duets de l'Amour". The floral wall decorations and a heart-shaped urn indicate the path of true love. However, a painting shows Cupid shooting at a pair of cooing doves with a blunderbuss, while 2 cats fight fiercely on the floor, perhaps foretelling a less harmonious future for the couple.

Above **Hermetic Divine Harmony (17th century) in Robert Fludd's *Ultriusque cosmi historia*.**

Above **The dynamics of creative number 3 is shown in the central symbol of 3 colours on this Japanese ceremonial drum.**

Above **Note the hexagram artwork on this ancient kissar (Nubian lyre).**

Above **Geometry, proportion, music, harmony and beat all feature in complex dances between partners.**

Counterpoint and resonance

Counterpoint is the joining together of two or more different musical lines that complement each other, while at the same time maintaining their own identities. Resonance is the relationship between moving bodies whose cycles are locked into one another, the planets being the most obvious example.

16th-century counterpoint

> Music is the shorthand of emotion.
>
> LEO TOLSTOY,
> RUSSIAN WRITER
> (1828–1910)

Harmony of parts in the whole

Throughout his book, *The Power of Limits*, architect Gyorgy Döczi beautifully illustrates the appearance of musical harmonic proportions in all forms in Nature, art, religious artefacts, architecture and the human form. For example, his illustrations of leaf formations show the stem always centred on a Circle in the middle of concentric Circles.

Harmonious sharing, working well together, are the principles of the Life Force and the Spiral and are linked to the number 5 and the pentagram and pentagon. Döczi explains how the power of the Golden Section creates harmony because of its unique faculty to unite different parts of a whole; while each part maintains its own identity and blends into the larger pattern of a singular whole. In principle this is the case with each Being manifest, where all individuals are parts of the larger harmonious pattern.

Perfect fifth example

Major third example

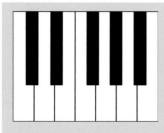

Octave example

Musical notes and harmony

Developed over many centuries, Western musical notes are derived from handwritten signs written above the words of medieval chants. First, notations defined the pitch of a note, then its duration. Originating from Pythagoras, the 7 notes of the octave (8 when you include the first note of the next octave) are divided into 12 semitones. Time signatures came next and finally the key. By the mid-18th century supplementary instructions were added, such as tempo, articulation, performing techniques and expressiveness.

Harmony exists because certain combinations of notes are deemed more pleasant than others. Longue-Higgins argues that a simple model can be used for the attribution of a musical key. He shows that every interval in music can be represented by a combination of 3 variables; octaves (7), perfect fifths (5) and major thirds (3). The listener attributes a key to music by selecting a region of this space. We can only discern 7 degrees in music with our ears. In a given key you can ignore the octaves and work with fifths and thirds (5 and 3).

Gregorian chant

Modern musical notation

Piano keyboard and the Fibonacci Series

Diatonic scales are the basis of Western musical tradition, visually evident on a piano keyboard as a 7-note octave-repeating musical scale comprising 5 whole steps and 2 half steps. The piano keyboard is also structured using the Fibonacci Series (see also page 53) harmonic progression. The 13-note chromatic octave comprises 8 white keys (whole tones) and 5 black keys (half-tones, sharps and flats) arranged in groups of 3s and 2s to make a full octave (including the start of the next octave).

The Fibonacci Series is a powerful type of harmonic progression and, by implication, the Spiralling Life Force of creation is a harmonic progression moving life through time. As shown in Chapter 8 (see pages 62–71) the Fibonacci Series is the secret to self-replicating growth in geometry and hence in Nature; an accumulative process that grows from within the original 0 and 1; the Point within the Void.

Left Piano keyboard octaves numbered from 1 to 8.

Solfeggio Frequencies

Solfeggio Frequencies is the term used to describe the 6 pure tonal notes once used to make up the ancient musical scale tones used in Gregorian chants, sung in harmony and in Latin. The chants and their tones were believed to pass on spiritual blessings and healing when sung in harmony. They vanished from use around AD 1050, to be unearthed by the late Dr Joseph Puleo (one of America's leading herbalists), as described in the book *Healing Codes for the Biological Apocalypse* by Dr Leonard Horowitz. Dr Puleo examined the Bible and noticed that there was a pattern of 6 repeating codes around a series of numbers 3, 6 and 9 in Genesis: 7, 12–83. When deciphered using the ancient Pythagorean method of reducing the verse numbers to their single digit integers, the codes revealed a series of 6 electromagnetic sound frequencies that correspond to the 6 tones of the ancient Solfeggio scale.

Each of the 6 Solfeggio Frequencies correspond to not only a note on the tonal scale, but to a cycle per second hertz (Hz) frequency number and qualities (defined by Horowitz):

UT 396 Hz Liberating guilt and fear
RE 417 Hz Undoing situations and facilitating change
MI 528 Hz Transformation and miracles (DNA repair)
FA 639 Hz Connecting/relationships
SOL 741 Hz Awakening intuition
LA 852 Hz Returning to spiritual order

According to Horowitz the 528 Hz frequency is known as the "528 Miracle" because it has the capacity to heal and repair DNA within the body and is the exact frequency that has been used by genetic biochemists. "The Hymn to Saint John the Baptist" has become known as the most inspirational hymn ever written and it features all Solfeggio notes.

> Without music, life would be a mistake.
>
> FRIEDRICH NIETZSCHE, GERMAN PHILOSOPHER, POET, COMPOSER AND CLASSICAL PHILOLOGIST (1844–1900)

Illustration of music ratios

1:1 = unison
1:2 = diapason, also known as an "octave", the same string 8 intervals of the scale, a rectangle of two equal Squares.
2:3 = diapente, the 5th, which approximates to the Phi Ratio
3:4 = diatessaron, the 4th equates to the 3:4 proportion of Pythagorean Triangle.

Above **The Phi Ratio in a pentagram.**

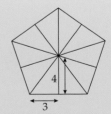

Above **Pythagorean Triangle in the pentagon.**

Above **A drawing of how mathematician Nathaniel Bowditch would have studied the harmonic motion in 1815.**

Locking rhythms and harmony

We are all locked into one another as rhythmic, harmonious wave patterns on a cosmic scale. Rhythmic vibration is essentially harmonious; sharing is universally present across all the planes of Being. In Chapter 15 we saw many examples of vibration, oscillation, pulsation and repeating cycles in media such as the water of oceans and winds of the air moving over the surface of the Earth. We are locked into these natural rhythms, orchestrated together as one harmonious symphony on a cosmic scale. Vibrating energy within the Void is the nature of Being, ethereal and matter (physical), as symbolized by the Triangle. We find it in light, sound, colour, plant-growth patterns, tides and our own biorhythms. Our heartbeats, sleep patterns and brain waves are all part of our inner clock – as well as allowing us to register the apparent passage of physical time they also place us in the cosmic clock. Many individuals can feel the vibrations of constant change that permeate reality, strumming the Grids of the Matrix of Space-Time-Being.

"Music is the pleasure the human mind experiences from counting without being aware that it is counting."

GOTTFRIED LEIBNIZ, GERMAN PHILOSOPHER
AND MATHEMATICIAN (1646–1716)

Galactic morphology

The diagram below shows the "tuning fork" of the Hubble system (1926) devised by American astronomer Edwin Hubble for classifying galactic morphology (the study of structure used by astronomers to divide galaxies into groups based on appearance). Though considered too simplistic by modern standards, the basic ideas still hold true.

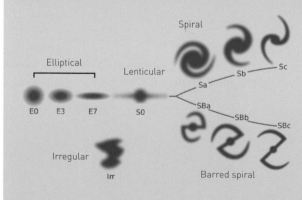

KEY	
E0–E7	Elliptic galaxies, progressively more eccentric
S0	Lenticular galaxies
Sa, Sb, Sc	Spiral galaxies
SBa, SBb, SBc	Barred spiral galaxies
Irr	Irregular galaxies

Lissajous Curve and knots

In mathematics, a Lissajous Curve (Lissajous figure or Bowditch Curve) is the graph of the system of parametric equations, which describes complex harmonic motion. This family of curves was investigated by mathematician Nathaniel Bowditch in 1815 and later in more detail by Jules Antoine Lissajous in 1857. Visually they resemble three-dimensional knots and are highly symmetrical.

A knot is a complicated concept in mathematics, where it is a closed curve in 3 dimensions that cannot be turned into a ring without cutting it. The ring (or Circle) is unique, whereas there are many different knots. This field of mathematics is used in String Theory and M Theory models to explain multiple dimensions and parallel Universes.

Lissajous 8_{21} knot

Lissajous square knot

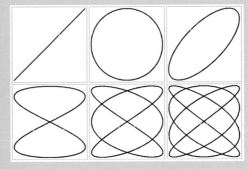

Formation of the Lissajous knot

Earth-Moon tone relationship

The geometry of the "squared circle" (see page 31) relationship reflects the Earth-Moon relationship. The large Circle of the Moon's path and the Circle circumference of the Earth come close to the ratio of 9:8, the basic interval of the tone of the musical scale.

Our body, soul and celestial harmony

Pythagoras believed that music was the sound of mathematics and laws of the Universe. Plato developed Pythagoras' idea to form the basis of a cosmological plan in which the harmonies of music, mathematics and celestial movements were all linked. Our ancient ancestors understood this link when they used the proportions of the human body and other natural phenomena for measuring distances to break up space.

Measurements of the human body tie us to the Earth and are the source of our traditional inches and feet measurements. For example, six feet becomes the fathom, or a man's outstretched arms. It is in using the ancient measures that we find human proportions that directly relate to measurements of the Earth, cosmic distances and these measures also link patterns found throughout Nature. Such patterns cannot be found using metric systems because they are based on logic alone rather than measurements provided by Nature, which require intuitive recognition and application.

Above "Saint Cecilia" by Raphael (16th century). St Cecilia was the patron saint of musicians and Church music because, as she was dying, she sang to God.

In addition to celestial harmony (*musica mundana*) Pythagorean musical theory identified two other varieties of music; the sound of instruments (*musica instrumentalis*) and the continuous unheard sound our bodies make due to the resonance of the body with the soul (*musica humana*). We appreciate music because the rhythms of our body and soul resonate with the harmony of the celestial realm, linking our physical and spiritual parts of Being through the geometric vibrations of sound in time.

Physical and psychological effects of sound

Sound affects the very structure of our Being as the waves ripple through us continually. No wonder the sounds of Nature, our voice and music hold such power over us. We sense the sounds of Nature, from the invigorating roar of the ocean, to the calming patter of raindrops. Similarly, man-made sounds cause positive and negative emotions, ranging from the pleasures of music to the dulling din of traffic. Our voice is an incredible tool – used to calm, excite and elicit fear, express hatred and love. Research has shown how music can affect the physical development of infants, alleviate depression, improve memory, reduce anxiety and aid concentration. Music, language, chanting and singing are composed so that sound is a tool that can be used to affect, even alter, our state of Being.

"There is geometry in the humming of the strings, there is music in the spacing of the spheres."

PYTHAGORAS, GREEK PHILOSOPHER
AND MATHEMATICIAN (C570–C495 BC)

Above **A page from** *A New Theory of the Science of the Music Tones* by Zhu Zaiyu (1536–1611) illustrating some of the geometric structure behind the organization of Chinese musical tones.

Above **Chinese musicians – note the symbolism of 3 on the motif.**

Chinese music and Universal harmony

Traditionally the Chinese believe that sound influences the harmony of the Universe and one of the more important duties of the first emperor of each dynasty was to find the new dynasty's true standard of pitch. For several thousand years the philosophy of Confucius dominated Chinese music, which was viewed as a means of calming the passions, dispelling unrest and purifying thoughts rather than entertainment. Melody and tone colours are important expressive features and emphasis is given to the proper articulation and inflection of each musical tone. Most music is based on the 5-tone, pentatonic scale. 5 is the number of harmony, sharing and structure of form. The 7-tone, heptatonic scale is also used.

African music

Music, especially drumming, has a sacred place in African cultures since sound is thought to be one of the primary means by which deities and humans impose order on the Universe. In addition, West African drummers play a crucial role in possession-trance ceremonies. The drummer knows specific rhythms of particular gods and is responsible for regulating the flow of supernatural power in rituals. The drummer strikes the drumhead to produce varying pitches known as "tone colours" and repetition is an organizing principle. Usually there is a chorus with a repeated refrain and a solo allows more freedom for improvisation. A variety of interlocking vocal patterns is also used. In traditional Zulu music, voices enter a cycle, overlapping in a complex, shifting pattern. Songs and dances are the medium for the transmission of knowledge and values, in the marking of traditional rites and celebration of events.

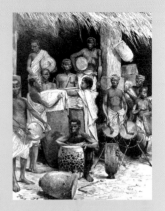

The Indian Raga

In India there are two groups of music, one for entertainment and the other to accompany life cycles and rites. Traditional music has been eroded by the music found in Indian films, but Indian classical music has been left largely untouched.

Classical music derives its inspiration from *bhakti* (devotional) movements. There is a melody line based on one of several hundred traditional matrices called "raga", which are written in terms of their emotional connotations, moods, colours, performance times and deities. In free time the melodic features are explored gradually in their natural rhythm, whereas in measured time a repeating number of units, the measure, is used to form a cycle. Within the cycle specific points receive different degrees of stress.

Above **Melakarta is a collection of fundamental ragas in South Indian classical music. The Katapayadi Sankhya, illustrated here, is a mathematical method to establish the number of a Melakarta Raga from the first two syllables of the name of the raga (by Basavaraj Talwar, Mysore/Bangalore, India).**

Part IV
LET THERE BE LIGHT

Anyone who has common sense will remember that the bewilderments of the eyes are of two kinds, and arise from two causes, either from coming out of the light or from going into the light, which is true of the mind's eye, quite as much as of the bodily eye; and he who remembers this when he sees any one whose vision is perplexed and weak, will not be too ready to laugh; he will first ask whether that soul of man has come out of the brighter light, and is unable to see because unaccustomed to the dark, or having turned from darkness to the day is dazzled by excess of light.

PLATO, GREEK PHILOSOPHER
AND MATHEMATICIAN (C424 BC–C348 BC)

Part IV – Let There Be Light – reveals how light can be considered as an essential dynamic in your perception of reality. Learn about the nature of visible light and mysterious dark matter, the importance of energy and Einstein's theories. You will discover why the colours contained within light and the correspondence between sound and light are important.

18 Light Beings

There are two ways of spreading light: to be the candle or the mirror that reflects it.

EDITH WHARTON, AMERICAN NOVELIST (1862–1937)

"GOD COMMANDED THE LIGHT INTO BEING through the medium of sound." Many belief systems describe the central role of light in creation. This part of the book examines light's role in detail. If sound contains and carries the geometric code, how does it influence light so that Being becomes? Let us start with understanding what light is and how it behaves. Light is a mysterious form of energy, with many "invisible" aspects. Light can travel through the vacuum of space, but no one actually knows what it is; only that it behaves both like particles and waves.

Incandescent light sources

Incandescent sources of light originate from sources of heat, like sunlight or a flame. Hot atoms collide and in the process transfer some energy to electrons, boosting them to higher energy levels. As the electrons release this energy they emit photons (see below). As an example, candlelight results from agitated atoms of soot in the hot flame. All incandescent sources of light have a broad electromagnetic spectrum. A good example is sunlight, resulting from nuclear reactions within the Sun. The light from the Sun is made from electromagnetic waves, of which only one per cent reaches the Earth's surface, since water vapour and ozone absorb much of this radiant energy. Moonlight is sunlight reflected off the surface of the Moon with a similar spectrum to the Sun, with a million times lower intensity, while starlight is a thousand times less intense.

Above **Light refraction through water makes the spoon look bent.**

Above **A flame on earth and a flame as it looks when burning in zero gravity.**

Light as particles and waves

Light changes its speed at the boundaries of different materials – this is called "refraction". When light refracts it exhibits wave patterns called "electromagnetic waves". When a wave travels in its electromagnetic field it must move at each of the finite number of Points, in every small part of space. In 1900 physicist Max Planck, regarded as the founder of the quantum theory, proposed the existence of finite packets of energy that Einstein named *quanta* and which are now known as "photons". Photons cannot be subdivided, have no mass or charge, but they do have energy and momentum. Unlike conventional particles, such as dust particles, photons are not limited to a specific volume in space or time.

Above **Light takes different forms: Sunlight, Moonlight, starlight, and light produced from extreme heat and burning in chemical processes, such as in fireworks and volcanoes.**

Light waves

Around 1870 James Clerk Maxwell, Scottish physicist and mathematician, showed that not only electricity and magnetism behaved according to wave equations. He discovered how interactions of electricity and magnetism travelled at the speed of light and produced light. Since then physicists have realized how we exist in a Universe of waves.

There is a magnetic field of positive and negative charges, constantly vibrating and producing electromagnetic waves. This excitation of static magnetism is a form of energy inherent in the force fields of all atoms. Each has a different wavelength and speed of vibration, which together form the discrete electromagnetic spectrum. Light created by atomic collisions in which the energy states of the atoms or molecules are too low to create visible light will create infrared light and at even lower energy states, radio frequencies. Light created by atomic collisions in which the energy states are extremely high will create X-ray and gamma ray electromagnetism.

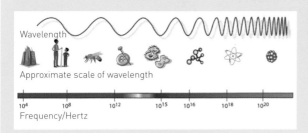

Left **The positions of frequency rates are shown on this very small section of the electromagnetic spectrum (NASA).**

All the darkness in the world cannot extinguish the light of a single candle.

St Francis of Assisi, Italian monk and saint (1181–1226)

Above **A light sculpture illustrating how fire acts in space due to the ventilation flow in microgravity. Blue areas are caused by chemiluminescence and the white, yellow and orange regions are due to glowing soot within the flame zone.**

Luminescent and fluorescent light

Luminescent light originates from sources other than heat, such as chemical or electrical processes. Fluorescent lighting is an example of a manufactured light form, as the light produced by televisions, for example.

Aurora Borealis and Aurora Australis (Northern and Southern Lights) are spectacular examples of naturally occurring luminescent light. Electrons in the solar wind sweeping out from the Sun become deflected in the Earth's magnetic field and dip into the upper atmosphere near the North and South magnetic Poles (see page 138). The electrons collide with atmospheric molecules, exciting the molecules' electrons and making them emit light. Chemiluminescence occurs in chemical reactions that produce excited molecules, which can then produce light. Some animals and plants make light in this way, for example fireflies. The pictures below show some examples of light, naturally occurring and manmade.

Northern Lights

A small portion of the Orion Nebula

Bitter Oyster fungus

Zooplankton beroidae

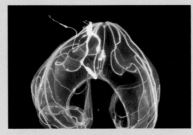

Deep-sea comb jelly

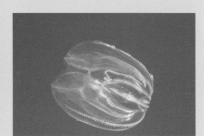

Leidy's Comb Jelly

Polyps on marine life

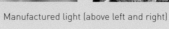

Manufactured light (above left and right)

Spherical nature of light and sound

Some scientists have argued that sound and light both travel as spheres. The activation of sound precedes the creation of light. When the force fields of atoms, or molecules, collide, there is a transfer of their periodicities, defined as the "phonon", or as sound. According to John Stuart Reid (Cymatics, see page 140) sonic sound bubbles expand at approximately 700 miles (1,126 km) per hour. A sonic bubble expands because the jostling air molecules cause friction and spread out at the atomic level, so that the sound can be made. Sound pressure rapidly decreases as a result of the initial energy in the sonic bubble being spread more and more thinly as the bubble's surface area expands. Jostling air molecules create friction, which theoretically then creates electromagnetic energy. It could be said that sound energy dissipates partly because of its conversion to electromagnetism. The sonic expansion generates an accompanying electromagnetic sphere that rushes away at 186,411 miles (300,000 kilometres) per second. In 1678 a Dutch scientist, Christian Huygens, proposed a successful theory of the light-wave motion in 3 dimensions. He suggested that light-wave peaks form surfaces like onion layers. In a vacuum or "uniform material" the surfaces are spherical.

Huygens' and Reid's theories for spherical light would also explain how light spreads away from a pinhole, rather than going in one straight line through the hole and why edges are blurred into shadows. While the energy in the sonic bubble drops rapidly with distance, the electromagnetic sphere travels relatively unhindered, through the atmosphere, to outer space. For example, starlight will travel virtually forever unless it meets with dense matter, where it will be absorbed and partially reflected.

Top and above **Light spreading away from its source – the Sun.**

Light cone

A light cone is the path that light would take when coming from a single event (in a Point and Moment in space-time, at the Centre of the Square and Circle) and then travelling in all directions. This diagram (right) shows 2 spatial dimensions and time as a vertical line. The top of the event is the future and the bottom is the future in reverse (or past). In 2 dimensions radial Circles (the Holon) of light would move away from the Point in the future, while contracting back in the past cone. Since there are 3 dimensions of space the light actually forms an expanding or contracting sphere.

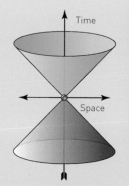

Above **A metaphorical visual representation of the concept of the expanding or contracting sphere.**

> Nature and Nature's laws lay hid in night./ God said, "Let Newton be!" and all was light.
>
> ALEXANDER POPE, ENGLISH POET AND SATIRIST (1688–1744)

Does sound create light?

Theoretically, maintains John Stuart Reid, there are two component frequencies of oscillation in the sound-generated electromagnetic sphere. The first frequency is that of light created by the collisions. This light oscillation is more than likely created by the innate sound periodicities of the colliding atoms or molecules. As it is predicted that sound always has an electromagnetic (light) component, then the frequencies of these components are either in the radio spectrum or in the infrared band, except where the sound pressure levels are extremely high. In these cases, sound would create visible light.

Light and waves of probability

The contradiction of light's behaviour as waves and particles led to the formulation of a quantum theory to try and explain how at a subatomic level matter does not exist with any certainty at definite places, but shows "tendencies" to occur. These "tendencies" are expressed as probabilities of interconnections. Light acts as a particle and a wave at the same time because light is not a real wave, like a sound wave in the air, but a wave of probability. As such they hold the potential to "become". This becoming is not an independent form, not a separate thing, but an event that occurs in Space-Time-Being. It contributes toward explaining why matter is essentially a highly complex web of relations between parts of the whole rather than isolated building blocks. Also, these relations can only occur to shape the many forms of matter at exactly the right temperature.

What we see as forms of Being, such as planets or plants, are all made of waves and processes that pass by. For example, we watch waves crashing on the shore, but what we actually see is energy travelling along to fall away (see pages 133–141). The wave itself is the Mind following the Eye that sees adjacent parts of the wave rising and falling. Matter's nature comprises wave after wave of energy arriving and departing from a Central location, which may itself be moving. This central location is the electron. Therefore every particle has its own characteristic wavelength and energy.

Abell 1689

Dark matter in the galaxy cluster is mapped by plotting the arcs produced by the light from background galaxies that is warped by the foreground galaxy cluster's gravitational field. Dark matter cannot be photographed, but its distribution is shown in the blue overlay. Images like these are being used to better understand the nature of dark energy, a pressure that is accelerating the expansion of the Universe.

Left The inner region of Abell 1689 (an immense cluster of galaxies located 2.2 billion light-years away). The picture was taken in 2002 using Hubble's Advanced Camera for Surveys (NASA).

Dark matter and gravity

Astronomers have long suspected the existence of the invisible substance called "dark matter" as the source of additional gravity that holds together galaxy clusters. Such clusters would fly apart if they relied only on the gravity from their visible stars. Although astronomers don't know what dark matter is made of, they imagine that it is a type of elementary particle that pervades the Universe. Dark matter neither emits nor scatters light, or other electromagnetic radiation.

Complementing the theory of light waves of probability is a theory that dark matter is moulded by contours in space amassed by gravity, whereas visible matter is buffeted by the energy of photons (light) and attracted to the structures made by dark matter. As visible matter collects it forms stars, galaxies and so on.

Gravity holding forms together

According to Bruce Cathie, since gravity and the speed of light are harmonic reciprocals of each other, as gravity increases the speed of light decreases, and vice versa. This supports the belief that as we are pulled to the physical plane and held there by the central force of gravity the vibrations of light appear to "become solid". Light is not a speed, it is an acceleration or deceleration according to geometric position. Because of relativity, objects' positions with regard to each other only *appear* to be constant. As gravity and light varies then all the physical processes vary in direct ratio, including the instruments involved in measuring objects. Every reading will appear constant, since time also alters in relation to gravity and light.

Because light behaves as waves and particles it vibrates at different rates and in the physical planes the light apparently solidifies or "crystallizes" in material form. Gravity is a state of tension that maintains the form. It is a ratio between geometric measures caused by the effects of bodies of matter relative to each other in space. No actual lines, just tension between Duals, facilitate ratios, measurement and patterns. Gravity also influences the causal relationships between Space-Time events, which are specific Moments, or Points in Space-Time. Gravity is the only physical quality that has this effect. Indeed, Einstein proved that gravity is different from the rest of physics. Matter is drawn toward the Centre of a Spiral due to the pull exerted by a gravitational field. In this way the millions of Spiral gravitational fields caused by the mini-vortices of each atom combine as one unique unified field for a form, such as a planet or a human body. When the Spiral motions cease then matter vanishes, reverting to the "primary substance".

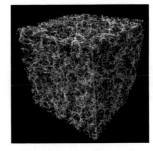

Above **This cube represents a slice of the spiderweb-like structure of the Universe, popularly called the "cosmic web", made largely of dark matter found in the space between galaxies. Dark energy is a theoretical form of energy that infuses all of space and is inclined to increase the rate of expansion of the Universe (NASA, ESA, and E. Hallman, University of Colorado, Boulder).**

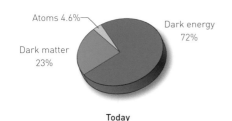

Atoms 4.6%
Dark energy 72%
Dark matter 23%

Today

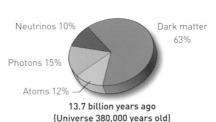

Neutrinos 10%
Dark matter 63%
Photons 15%
Atoms 12%

13.7 billion years ago
(Universe 380,000 years old)

Left **The approximate distribution of dark matter and dark energy in the Universe (NASA).**

> Energy
> is eternal
> delight.
>
> WILLIAM BLAKE,
> ENGLISH POET, PAINTER
> AND PRINTMAKER
> (1757–1827)

The Energy Grid

In his Theory of Relativity Einstein argued that time and space are two aspects of the same thing, as are matter and energy. Even when an object is still at rest it has energy stored in its mass. Einstein believed that M (mass) could be replaced with a term denoting wave form, representing pure energy; a form of energy that represents the existence of any physical manifestation. In *The Energy Grid* and *The Bridge to Infinity* author Bruce Cathie replaces M with a term denoting light waves. The revised Einstein equation is an equation expressed in terms of light. In essence this implies that the whole of existence comprises seen and unseen light waves. The Cathie system of "light harmonics" is based on the geometry of the Circle, the circumference of which is divided into 6 equal parts and a number 6 is used to shift values up and down on his harmonic scale. Number 6 is part of the Grid of Life (see Chapter 4, pages 34–41) and can be drawn as the Second Level Triangle. Base 10 is also a fundamental part of the harmonic process in Cathie's calculations, and 10 is the Third Level Triangle (or Tetraktys).

Moulding matter out of light

Geometric combinations mould matter out of resonating, interlocking light-based waves. Matter and anti-matter are formed from the same Spiralling wave motions in space, where matter is a positive state and anti-matter a negative one. Each Spiral of 360 degrees is a pulse. The relative motion of nucleus and electrons through space gives an illusion of a Circular motion, since its period through anti-matter is not detectable using physical instruments. The frequency rate between each pulse of matter creates the illusion of linear time and light in the particular position in space we are at any given moment.

Colin Wilson (English philosopher and writer) describes how if we turn through a full Circle we find ourselves in our original direction. However electrons' "axis of spin" can be tipped through 360 degrees when passed through some electromagnetic fields but then do not restore themselves to their original position. The electron has to be turned through another full Circle before it behaves as before. While we cannot distinguish between the two Circles the electron can. Wilson believes that this suggests that in the subatomic world a full Circle is not 360 degrees, but 720 degrees and that either we have lost half the degrees we ought to have or that there may be another dimension in the subatomic world.

The Prime Number Cross

In *God's Secret Formula*, chemist and mathematician Dr Peter Plichta made a few key observations that relate to the 3-fold nature of Being and its ties with the 4-Square law. He also noted that much of the structure of matter, social organizations, languages and culture are 3-fold in nature, indeed atoms have no more than 3 components. Furthermore, the Square law is anchored into the atom and hence all matter. Even empty space has a 4-fold structure, so the two-dimensional Square (4), and then the 4 quantum numbers of electrons must be a necessity. Atomic structure uses the numbers 3 and 4 and this principle can be extended up the scale of Being. For example, DNA consists entirely of 3 chemical compounds: phosphoric acid, sugar and a base. These 3 items form a chain in which a base is connected to each sugar molecule. This base may be one of 4 bases (thiamine, adenine, cytosine and guanine).

Plichta discovered a pattern in the prime number series that has a direct correlation with the organization of atoms. In essence it is an atomic model: $1 + 2 + 3 = 6$ and $1 \times 2 \times 3 = 6$. That both the sum and product of 1, 2, 3 equal the same number is unique to these 3 numbers. So 6 is the Grid on which the prime and natural numbers are built.

Plichta constructed a model called the Prime Number Cross. He gave the model this name because of its shape when represented graphically as a series of concentric Circles, highlighting the importance of regarding numbers as Circular rather than linear (see page 59). For example, he notes that the Babylonians set the length of a second at 3,600th part of an hour – the product of the Pythagorean numbers 3, 4 and 5 squared – and one hour was the 24th part of a day = $1 \times 2 \times 3 \times 4$.

Plichta then turns his attention to the value of the speed of light. Though Einstein and other scientists calculated the absolute speed of light to be 2.9979, Einstein referred to the speed of light with a rounded-up value of 3×10 to power of $10 \times$ cm/s. Plichta speculates that this rounded value is the absolute value and something that is not just a matter of chance. He also reasoned that the decimal system cannot be a coincidence and measurements of length, weight and time could be identical with the absolute dimensions in which Nature is arranged.

> Come forth into the light of things,
> Let Nature be your teacher.
>
> WILLIAM WORDSWORTH,
> ENGLISH POET
> (1770–1850)

Origin of energy and the hexagon

Also of interest is the work of Jason D. Padgett, who investigated the origin of energy (2005) and his results are shown as the hexagram grid, which he calls the "Planck Lattice". This correlates with the Grid of Being within the Matrix of Space-Time-Being. Each side of every small equilateral Triangle in the Planck

Above **Sunlight showing us physical forms.**

Lattice represents a Planck Constant or a Planck Length. A Planck Length is the smallest piece of Space-Time that can be observed (or exist relative to us). As there are no measurements smaller than a Planck Constant you cannot have a distance that is 1.5 Planck Lengths. This means that when measuring slices of Space-Time you can only measure 1, or whole multiples of 1 Planck Lengths. Any Point of Space-Time that exists must be exactly 1, or multiples of 1, Planck Length from all other points. This automatically creates the hexagonal grid-like structure of Space-Time. Because there are no fractions the shape that arises is a two-dimensional hexagon interacting with other hexagons, creating larger and larger hexagons; the Grid of Being. If you view this three-dimensionally you will see that it automatically creates three-dimensional cubes that create larger and larger cubes (see page 114).

When we are looking at a point of Space-Time, or Planck Constant, it "vibrates" at the speed of light and we cannot define its position and velocity perfectly. Imagine the Grid of Being, or Planck Lattice, vibrating at the speed of light from the Centre. The vibrations move outward and collide with each other at specific points. According to Jason D. Padgett, Einstein's $E = MC^2$ is energy = mass x the speed of light squared.

Above **A pious man's soul from his deathbed to his judgement before God where his true "Being" is revealed by God's light as depicted in** *The Dream of Gerontius* **by Stella Langdale.**

Einstein's energy and Planck's energy

Vibration, energy and mass that we experience as a physical entity are explained in a mathematical equation that tells us that they are made of light. Energy = the total Planck Constants in the vibration (which is their mass) times the speed of light (how fast the Planck vibrates), squared (as the Planck Constants collide head on at the speed of light). Max Planck, who discovered and named the Planck Constant, won the Nobel Prize in physics and discovered that Energy = hf. In words: energy equals the Planck Constant (h) multiplied by frequency (f). Now Plank Energy = hf and Einstein's Energy = mc^2, therefore hf = mc^2.

Combining these two equations and seeing the shape of Space-Time vibrating at the speed of light is the key to understanding where energy comes from. In this way the Gateway to the Heavens model is actually made of energy and therefore light. What this model does is give energy the ability to shape reality, to hold and sustain itself as space, time and myriad Beings of light.

Above **Sunlight being reflected is in essence a light source revealing itself to itself. It is also symbolic of the nature of physical light (body) and ethereal light (spirit).**

Light creating light

Light travels, then meets a material and in turn interacts with the atoms in the material. In transparent materials the electrons vibrate while the light source is there, taking energy and then returning it. Other materials either absorb or reflect light. If absorbed, the energy of the oscillating electrons does not return to the light, but increases the motion of the atoms and this makes the material heat up. When reflected the atoms re-radiate the light and cancel out the original wave. In effect, the light reveals the matter that is part of physical Being. *Light reveals Being and light is generated by Being.*

Light of Being, Being in light

Whenever the question is asked as to what Beings are, Beings as such are "in sight". Metaphysical thought owes this sight to the light of Being. A Point within every Being exists which is the source and originator of the light, allowing every Being to be "in sight". However the nature of Being is defined, whether as spirit, or as matter and force, or as becoming, as an idea, via will or as energy, Beings still appear in the light of Being. Being has entered into the light so that it can be revealed.

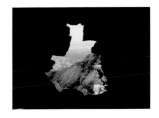

Above **Emerging from the darkness light reveals Being.**

Physical light and physical life

Without the light of the electromagnetic spectrum, physical life as we know it would disappear. Light provides energy for plant growth and is stored in chlorophyll for use in the process of photosynthesis (the process where plants convert carbon dioxide from the air into organic compounds using the energy from sunlight).

Petroleum, gas and coal are made of the remains of plants that lived millions of years ago. When these fuels are burnt they release the energy stored by the plants; energy converted from sunlight. In effect we eat and absorb light to survive as physical Beings.

The Life Force and Universal Energy Field

Many spiritual texts, particularly those on healing, refer to the Universal Energy Field (UEF), which is considered to exist throughout space and all things, animate and inanimate, and which also connects everything together. It is suggested that the density of the UEF diminishes according to the distance from its source and that the UEF follows the law of harmonic inductance and sympathetic resonance, hence vibrating at the same frequency as an object it is close to.

The UEF is held to be highly organized geometrically, exhibiting harmonic attributes, pulsating points of light, Spirals, webs, sparks and clouds. Constructively the UEF builds and organizes by providing the blueprint across all planes of Being and that on which matter forms. Apparently, any changes in the physical Plane are preceded by a change in the UEF. The UEF, considered to be in all forms of Being, has delicate strands connecting everything. It too can be thought of as just such an electromagnetic force, with positive and negative charges. The charge is minute and electricity powers electronic equipment so it powers all forms of Being.

> More light! Give me more light!
>
> JOHANN WOLFGANG VON GOETHE, GERMAN DRAMATIST, NOVELIST, POET AND SCIENTIST
> (1749–1832)

19 Colours of light

I never saw an ugly thing in my life: for let the form of an object be what it may – light, shade, and perspective will always make it beautiful.

JOHN CONSTABLE,
ENGLISH ROMANTIC PAINTER (1776–1837)

Top and above **Splitting white into spectrum light using a Triangular prism.**

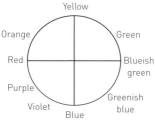

Above **The Centre of the additive colour Circle is white. Opposite colours are complementary. Note the geometric structure of this model.**

ACCORDING TO THE MYTHOLOGY OF NUMEROUS CULTURES the origins of the Universe are dark and watery, but the actions of light on water give us the rainbow spectrum of colours that bring vibrancy to Being. From the darkness light poured forth and the light gave birth to the colours. When colours are mixed they may lose their individuality, but they create a wonderfully diverse palette from which we can paint the images of the Universe to excite our emotions.

Additive and subtractive colours

Direct light emanates from a source, such as the Sun, a light or a star. Pure light consists of a 3-fold, Triangular weave of 3 primary colours – red, green and blue/violet. These primary colours are additive, braiding together to recompose whole light. Pigment colours, as in paint, are seen as reflected light. Their 3 primary colours are red, blue and yellow. These are known as subtractive colours, as they remain after direct light has been absorbed into, or subtracted from, the surface.

Direct light colours add to make white light and pigment colours subtract toward black, the absence of colour. In effect the reflected colours of objects show us what the colour is not because its true colour has been absorbed. All we see are the reflected remains. The colours of pure white light (spirit: heaven) and colours of pigment (matter: Earth) build in opposite directions.

The colour spectrum

The majority of cultures identify a small group of 7 definite colours, as seen in the rainbow. They fall within the range of frequencies in the electromagnetic spectrum that are visible to the human eye. Though the differences between the 7 colours tend to be exaggerated, they are generally accepted as being red, orange, yellow, green, blue, indigo and violet. As we have seen (see page 161), the visible light spectrum is a miniscule part of the vast electromagnetic spectrum. The longest wavelength most humans can see is deep red at 700 nanometre (nm) and the shortest wavelength is deep blue/violet at 400nm. Human eyes respond best to green light at 550nm, which is also the brightest colour in sunlight at the Earth's surface and sits in the middle of the spectrum.

Arranging colours as a wheel and Triangle
Most colour wheels are based on 3 primary colours, 3 secondary colours and the 6 intermediates formed by mixing a primary with a secondary, known as tertiary colours, for a total of 12 main divisions. Other colour wheels are based on the 4 opponent colours and may have 4 or 8 main colours.

Above **Light as a spectrum.**

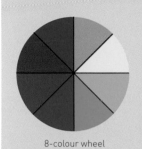

8-colour wheel

Sir Isaac Newton's basic colour wheel

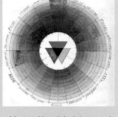

Moses Harris' elaborated colour wheel

Above **Lichtenberg colour Triangle by Tobias Mayer (1775). A colour triangle is an arrangement of colours within a Triangle, based on the additive combination of 3 primary colours at its corners.**

Colour Triangles and wheels

A colour Triangle is an arrangement of colours within a Triangle, based on the additive combination of 3 primary colours at its corners. Sir Isaac Newton is credited with being the first person to arrange spectral colours in a Circle, in 1666. Then in the 18th century, Moses Harris elaborated on the Newtonian colour wheel by adding various shades of those colours as well. His goal was to establish the complete world of colours resulting from mixture of the 3, among themselves and together with white and black. Meanwhile Tobias Meyer arranged colours within a Triangle, comprising of additive combinations of 3 primary qualities placed at its corners.

Adding dimensions to colour

Imagine the Newtonian wheel as a one-dimensional representation of colour and Harris' wheel with degrees of shade introduced to make it two-dimensional. In 1810 Philipp Runge, a Romantic German painter and draughtsman, portrayed colour as a three-dimensional sphere as he recognized that luminosity and saturation were unique quantities to be plotted on separate axes with white and black forming the opposing poles (see page 172). His dimensions are value (for example darker, paler), saturation (for example rich, pastel) and hue (for example orange, purple, green). Runge's system of hue, saturation and value is the most common, practical model in use since it reflects how we talk about colour. What is wonderful about these colour wheels is the Circular feature of colours mixing and blending to create varying hues and shades, as is their basis being Centred around a black or white Point of "gravity".

Additive and subtractive colours

When additive and subtractive colours are combined, as in the Ostwald colour solid below, you can envisage the spreading effect of light from a central source and its return to the dark that absorbs light. Imagine this as spinning as well, such that the colours blend, as they would to create the many hues of Nature.

Above **Physicist and mathematician, James Clerk Maxwell used spinning discs (1861) to compare a variable mixture of 3 primary colours with a separate sample colour by observing the spinning "colour top."** He provided the theoretical framework for practical colour photography.

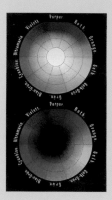

Left **Additive and subtractive colour wheels including tone by Bezold Farbentafel (1874).** In a light colour Circle, the Centre is white or grey, indicating a mixture of different wavelengths of light. In a paint, or subtractive, colour wheel, the Centre of gravity is usually black, representing all colours of light being absorbed.

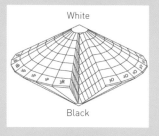

The Wilhelm Ostwald colour solid (1916)

Colour cones and light cones

Alvy Ray Smith, American engineer and pioneer in computer graphics, produced SuperPaint for Xerox and in 1978 formalized the HSV (hue, saturation, value) model – one of the earliest computer graphics applications. While computer screens actually function by mixing red (R), green (G) and blue (B) light, RGB is a technical system not an artistic model. HSL (hue, saturation, lightness)and HSV are the two most common cylindrical-coordinate representations of colour points in an RGB colour model. What is interesting about the HSV system is that it is represented in a cone shape, similar to the light cone, rather than as a sphere.

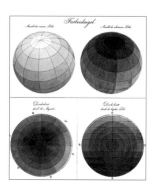

Above **Philipp Runge's colour as a three-dimensional sphere, adding luminosity and saturation on separate axes with white and black forming the 2 opposing poles.**

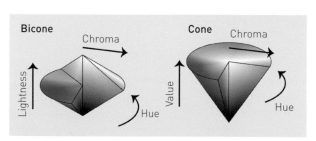

Left **Two cylindrical representations of the RGB system (© 2010 Jacob Rus).**

Seeing colours the human way

A key that links light to the Triangle of Being and our 3-dimensional visual system lies within the structure of the physical eye. Cone cells, or cones, are photoreceptor cells in the retina of the eye registering 3 separate pieces of information that are weighed and combined to produce our final sensation of colour. They work best in relatively bright light, while rod cells work better in dim light.

There are 3 kinds of cones. These are L (long) for long-wavelength light, peaking at a greenish yellow colour, M (medium) for medium-wavelength light, peaking at a green colour and S (short) for short-wavelength light of a bluish colour.

We can view objects under a variety of different lighting conditions. The human visual system adapts to these differences by chromatic adaptation, which is easily misled, resulting in common optical illusions relating to colour.

In normal human vision, wavelengths of between about 400 nm and 700 nm are represented by an incomplete Circle, with the longer wavelengths equating to the red end of the spectrum and violet at the other end. Humans perceive long-wavelength red and short-wavelength violet to be more alike than any of the other colours of the spectrum, even though red and violet are at opposite ends of it. This apparent similarity makes us join the two ends of the spectrum. Sometimes a gap is added to represent colours that have no unique spectral frequency. These extra-spectral colours, the purples, form from an additive mixture of colours from the ends of the spectrum.

Black, shadow and depth

The simplest human languages only describe two colours, black and white; the Duals that can combine in many shades and are fundamental to the principles of duality as represented by the T'ai Chi symbol. Both are necessary to convey the relative light and darkness of a daily cycle, temperature variation and colours of Nature in the seasons and the dynamics of our senses and emotions. Colours of light are additive, blending to white, while pigment colours of the Earth are subtractive, merging back to black. Each type is a reflection of the spirit and matter.

Pure, radiant, hot white light is the source of life and all colours, revealing Beings so that they can reveal all of Nature. White is illumination-spreading light so that we can see and gather knowledge. Cold black is the absence of light; the night and silence. The absence of illumination is a lack of knowledge and source of ignorance. But without black there would be no subtle variation in tone to enable shadow, the perception of depth and volume, degrees of white and black.

> *"It was one of those March days when the sun shines hot and the wind blows cold: when it is summer in the light, and winter in the shade."*
>
> CHARLES DICKENS, ENGLISH NOVELIST (1812–1870)

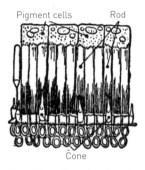

Above **Location of rod and cone cells.**

Above **Symbolically letting light in through the "window" of our eyes lets in the knowledge of what is outside. Tones of black and white reveal the body of Being in space; its volume.**

Noticing and naming colour

In cross-cultural studies of language, after black and white, red features as the next-most-common colour word. Red is followed by green (not surprisingly, as it is a colour we find in abundance in vegetation), then comes yellow and blue, followed by a distinct category for brown and finally purple, pink, orange and grey.

> ### Tetrachromacy
> Tetrachromacy is the condition of possessing 4 different types of cone cells in the eye for conveying colour information. Living organisms with tetrachromacy have to be a mixture of at least 4 different primary colours. For example, the 4 pigments in a bird's cones extend the range of colour vision into the ultraviolet. Fish, amphibians, reptiles, arachnids and insects are all believed to have tetrachromacy.

Bringing out the form of Being

What do we see? Not lines, but forms structured by geometry and coloured by light. Being is revealed by light. As light impacts with the surface of matter, light is absorbed and reflected into our eyes and viewed. Colour and tone reveal the form of Being. Indian gem therapists believe that everything is composed of the 7 rays (the primeval forces of Nature) and it is through the combination of these that tangible forms are produced.

Psychological and physical effects of colour

Light not only reveals form of Being, it contributes another layer by adding vibrancy and dynamism to the form, which has an impact on the observer. Light literally colours, or flavours and enhances, our experience of life. By looking at forms of Being we are able to see their physical form and are affected

Below **Colour abounds in all aspects of life on Earth, enhancing experience for all its inhabitants.**

Colourful supernova
NASA's three great observatories – the Hubble Space Telescope, the Spitzer Space Telescope and the Chandra X-ray Observatory joined forces to probe the expanding remains of a supernova, called Kepler's Supernova Remnant. Each colour in this image represents a different region of the electromagnetic spectrum. By colour-coding their data and combining them with Hubble's visible-light view, astronomers are presenting a more complete picture of the supernova remnant.

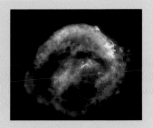

Supernova

psychologically, emotionally and physically by their colours. Colour surrounds us, nourishing our senses, but in the modern world, where we are constantly bombarded by colour, many people have lost touch with its meanings. Our minds, bodies and spirits are profoundly affected by colour.

Johann Wolfgang von Goethe's *Farbenlehre (Theory of Colours*, 1810) was the first study of the physiological effects of colour. He wrote that colour was a living entity of spiritual significance and stressed the importance of experiencing colour as a vital energy of life. His observations on the effect of opposed colours led him to a symmetric arrangement of his colour wheel. Goethe's work influenced philosopher Rudolf Steiner, who was also convinced that colours affected our emotions and body. Hence colour is an important theme in Steiner schools.

It is currently being scientifically proven that colour has a direct effect on us physiologically; on our moods, mental state, emotions and even health. Every colour spectrum of shades has an effect. The darker and denser the colour, the more physical it is. Lighter, pastel shades of the same colour have higher, faster vibrations. In this way, we can see, for example, how the blood red used to colour a heart symbol represents earthly love, while light pink vibrating at a higher level is related to softer, spiritual Love.

We see colour, feel it and absorb it. Each colour vibrates to a unique frequency and is visually distinctive. Our bodies are stimulated and energized by some colours; calmed and relaxed by others. What might be "good" for one person

> Colour in certain places has the great value of making the outlines and structural planes seem more energetic.
>
> Antoni Gaudí,
> Spanish architect
> and designer
> (1852–1926)

Colour in pattern-sharing
When 2 colours are used in a linear frieze using the 4 methods for repeating patterns (translation, reflection, rotation, glide reflection) they can be combined in only 7 different ways (see page 185). When there is a flat, two-dimensional surface (e.g. a carpet) there is more scope. There are now 17 ways of repeating a 2-colour pattern using the basic pattern-sharing methods; all known and exploited by the Ancient Egyptians for decorative purposes. Any one pattern can be combined with another pattern for an infinite number of possibilities and this is how simplicity allows diversity and complexity.

> There is not one blade of grass, there is no colour in this world that is not intended to make us rejoice.
>
> JOHN CALVIN,
> FRENCH THEOLOGIAN
> (1509–1564)

could be "bad" for another. Mentally and emotionally, colour works on a deep level, changing our mood and sense of well-being, as well as others' perception of us. Each colour also has its own taste and feeling. Research has shown that blind people can be affected by colours, frequently able to identify them via their fingertips. Red seems warm, rough and tingling, whereas blue seems smooth and cool, yet the measurable surface temperature of the colours is the same. Other sentient physical Beings are also affected by colour. In the animal and plant worlds colour can mean survival or extinction; it is used to attract, camouflage, ward off danger and send sexual signals.

Colour is intrinsic to life and as important to us as it is to plants and animals. It is as essential as food for health and well-being. Much effort has been applied to determining the use of shape and form in design, much less to the use of colour, but our innate sense of colour is no less important than our instinct for order and pattern. Colours are a gift of the Sun; without them the world would be uninspiring – they are a reflection of the emotions we experience through living.

Colouring our soul

As colour is the language of the soul, spiritually colour is of great significance. Down the ages humans have used colour in their rites and religions; from the saffron robes of Buddhist monks, the blue of the Virgin Mary's cloak in Christianity to the black and white worn to represent death, birth and renewal the world over.

Above **Colours are used to enhance the human form the world over. Without colours we lose a little of our humanity.**

Colour schemes

Though the colour wheel is used for illustrating additive colour mixture it is commonly associated with the concept of colour harmony. Adjacent colours on the colour wheel are seen to be not visually harmonious, while opposing colours complement each other, creating the best colour combinations. This understanding of complementary colours is the basis of modern colour schemes.

Above and top **Red and green (blue and yellow mixed).**

Above and top **Yellow and purple (blue and red mixed).**

Above and top **Blue and orange (yellow and red mixed).**

The symbolic qualities of colours

Colour symbolism varies slightly between different cultures, belief systems and even genders. However, there are many underlying similarities in the allocation of a colour to a symbolic meaning and purpose.

White Ultimate purity and innocence are white and cool. White is the light of day and being awake, it is a dispersed, radiant and available energy, also symbolic of perfection, illumination, truth, holiness, redemption and the triumph of spirit over flesh (which is why brides wear white). When white is associated with death, it symbolizes death of the physical body, making way for rebirth in new life. White is a colour of protection, bringing peace and comfort, helping to clear emotions. White gives a feeling of freedom and uncluttered openness, but too much can be cold and isolating. In Buddhism white means self-mastery and redemption, being led from bondage to the spiritual transformation of Nirvana. White represents loftier values as in, "I saw the light" and represents ultimate transcendent reality. In Christian belief white is the highest colour, representing the purified soul: joy, virginity, integrity, light and the holy life. In Hinduism white is symbolic of pure consciousness, or self-illumination, light and manifestation.

Black is associated with silence, the infinite, dark uncharted territories and mystery. In black there is no identity, only seclusion and a state of sleep at night. We cloak ourselves in black to hide from the world. It is the colour of transformation, since life needs to die to go into the blackness of death, to re-emerge in the light of life. Like the black hole absorbing and taking energy, black concentrates energy. It is the darkness of the night and north according to Native Americans and the Chinese. In Hinduism, black is Kali and corresponds to *tama*, the first of the 3 gunas, or states of Being when asleep. Buddhists regard black as a colour of bondage. According to alchemy black is absence of colour and physical life and in numerology black is represented by the number 8.

Grey acts as a shield from outside influence. Generally the colour is associated with fog, clouds and smoke. Since grey is neither black nor white it is the tone of evasion and non-commitment. Its more positive associations are independence, self-reliance and self-control.

Red has the slowest rate of vibration and is the densest, most physical of all the colours. Usually red is linked to the earth and to blood. Literally the colour of blood, it also has a stimulating action on our heart and circulation and red light raises the blood pressure. Our body system is fortified by red, it stimulates the adrenal glands, helping us to become strong and full of stamina. Red splits the ferric salt crystals into iron and salt and red corpuscles absorb iron, while the kidneys and skin eliminate the salt. Red is a "hot" masculine energy that is vital, empowering and stimulates action. Native Americans place red in the south, where the Sun is at its highest position in the sky.

Orange is a warm, happy colour. Linked to creativity, it liberates emotions and renews interest in life. It is a wonderful anti-depressant, lifting the spirits. Literally interpreted, orange promotes the assimilation of your life, or your ability to "digest it", vitality and the ability to deal with excesses. Orange symbolism is rooted in China and Japan. Citron is one of the three blessed fruits of China and the "fingered citron" is said to resemble the shape of Buddha's hand.

Yellow is also a warm colour – bright and uplifting. Yellow wavelengths of light stimulate the intellectual side of the brain, making you alert, clear-headed, decisive and able to remember things better. It is the colour of detachment and the power of judgement; the higher mental intellect of the brain and our rational, discerning aspect. Yellow also helps with the assimilation of new ideas and the ability to see different points of view. It builds self-confidence and encourages an optimistic attitude. It is also linked to the nervous system and the healing process. Native Americans allocate yellow to the west and the setting Sun. In Buddhism yellow is sacred and symbolic of renunciation and humility. The Chinese lunar hare is associated with yellow, the animal of all Moon deities and rebirth. For the Hindu yellow represents light, life, truth and immortality.

Green Known as the "fulcrum colour", green is the midway colour on the colour spectrum. Neutral in temperature, green brings physical equilibrium where positive and negative are balanced. Green has a strong kinship with Nature, helping us to connect and empathize with others and the natural world. Instinctively we seek green when under stress as it creates a feeling of relaxation, calm, space and it balances the emotions. Because of this green is one of the major healing colours. In Buddhism, vernal green is the colour of life and pale green the kingdom of death and everything pertaining to death. In Christianity, vernal green represents immortality and hope, the growth of the Holy Spirit within humans, life and spring. Green was the colour of the Trinity and the Epiphany in Medieval times. For Celts, green symbolized the Earth Goddess. Green marked the beginning of the Great Work for alchemists and was used in preparation for transmutation of base metals into gold.

Blue is a cool, calming colour. Like yellow, it is associated with the intellect, but it represents truth, inspiration, devotion and wisdom. Blue eliminates confusion so one can see with clarity and perceive integration. Darker blue, like the night, is soothing, calming and relaxing. Light and soft blues make us feel quiet and protected. As it opens up and slows down, blue creates an impression of space and expansion in an environment, like the vast sky. Blue is an excellent colour for meditation and is often used in places of healing. Blue is often associated with heaven, the sky and sky deities. Native Americans use blue to symbolize the sky and peace, placed in the east. For Buddhists blue is the coolness of the heavens above and the waters below. Christians use blue to symbolize the Virgin Mary, the Queen of Heaven and the mother of Christ, the representation of heavenly truth and eternity.

Turquoise is the last colour before the blue end of the colour spectrum. Comprising blue and green, turquoise can be biased to one or the other. It is used in colour therapy to strengthen resilience to overcome negativity (physical, emotional and spiritual) and boost the immune system.

Indigo comprises blue and violet. Referred to as the "vault of heaven on a Moonless night", indigo is associated with the right side of the brain, intuition and memory of our dreams.

Violet is used to denote regal stature, priestly power, authority on high, highest truth and penitence. In many spiritual traditions violet represents the last stage through which we must pass in order to become united with our inner Selves, so it can be used to lift the prepared human being into higher states of consciousness.

Magenta helps us to let go of the past, to release ideas, feelings and old patterns of behaviour that are no longer appropriate, so that we can evolve and grow. When magenta fades to pale pink it becomes the colour of spiritual love.

Purple is a psychic colour, also associated with the right side of the brain, therefore stimulating inspiration and imagination. Purple is also connected with artistic and musical impulses, mystery, sensitivity to beauty, spirituality and compassion. Christians use purple to represent God the Father.

Brown is symbolic of the physical earth and growth, since it is formed by the combination of colours derived from plants and stones when mixed on a palette. This very physical colour is strongly associated with the physical plane of Being; it is nurturing and brings a sense of stability, alleviating insecurity.

Silver is cool and frequently referred to as the colour of the female Moon, which is perpetually changing. It balances, harmonizes and is mentally cleansing.

Gold, like yellow, is warm and associated with the male Sun. It is related to all the Sun gods and those gods associated with harvest time. Therefore gold is related to abundance and power, higher ideals, wisdom and understanding. It can also help to resolve fear and uncertainty. In Christianity gold represents God as uncreated light and divine power. The Sun God Zeus is symbolized by a gold cord on which the Universe is said to hang and the "rope of Heaven" upon which all things are threaded.

20 Music and colour

Music is the arithmetic of sounds as optics is the
geometry of light.

CLAUDE DEBUSSY, FRENCH COMPOSER
(1862–1918)

Above **Conceptualization
of the rainbow colours and
shades of music.**

VIBRATIONS OF SOUND HAVE NO FORM themselves except as vibrations that must
move molecules if they are to be heard. Sound acts as a vehicle that contains
the geometric formula that works its magic, using vibration to structure forms
of light Beings. As Bruce Cathie maintains, the elements can be built simply by
bringing together combinations of lesser elements as long as they are in an environment of harmonically tuned sounds and correct temperature. Harmony is the
key word here, which we need to look at again.

Light and sound are not the same phenomenon as we have seen in the previous chapters, but they do share the vital essence of harmony. Eye and ear share
harmonious rhythm, with one registering it as sound and the other as colour.
Colours of light are moulded by sound. This is true "dinergy", as György Doczi
(see also page 151) terms it, an inseparable union of energies that form the
aspects of our Being and consciousness.

*"All human actions have one or more of these seven causes: chance, nature,
compulsions, habit, reason, passion, desire."*

ARISTOTLE, GREEK PHILOSOPHER
(384–322 BC)

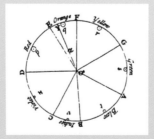

Above **Isaac Newton's
wheel of colour and musical
notes.**

Colour and music

Isaac Newton linked his colour wheel to the
white-note scale on the piano. Starting at D,
he divided his wheel into musical proportions
around the circumference as arcs from DE to
CD. Each segment was given a spectral
colour, starting from red at DE, through
orange, yellow, green, blue, indigo to violet.
The middle of the colours, or their "centres of
gravity", are shown as p, q, r, s, t, u, and x in
the diagram (see left). The Centre of the
Circle, at O, was presumed to be white.

Within us pitch, melody, loudness, rhythm and sound sources are all processed separately then reassembled and combined with an emotional and physical response. So the combined effects of sounds and colours of the Universe induce a wide variety of emotions and other physiological effects.

Sharing wave patterns

György Doczi, in his book *The Power of Limits*, correlates spectral colours (wavelength in Angstrom Units) with frequencies of notes (Hertz cycles per second) within one musical octave. Therefore colour and sound share the same wave patterns and same vibration rates.

Note	Colour	Note	Colour
G	Violet	D	Yellow
A	Indigo	E	Orange
B	Blue	F	Red
C	Green		

When the colours are put in a colour wheel and Triangles are drawn to link 3 notes (triple weaves) it is interesting to see that the note-to-colour combinations are found in Nature's harmonious colour patterns. Below are examples of colour combinations equating to musical chords.

Top row below Musical notes A:C:E = indigo-green-orange

Far left and left Musical notes G:B:D = violet-blue-yellow

Music and colour in Nature

In the myriad forms that are the product of Nature, we can see the harmonious interplay of colour and sound. Referring to Doczi's table (see page 181); the 3:4:5 Triangle turns out to be the tonic chord of A minor, or the notes A, C and E on the keyboard and indigo-green-orange on the colour wheel. Tonic chord G major, the notes G, B and D equate to violet-blue-yellow. We see these harmonious colour combinations replicated in Nature, especially in flowers.

3:4:5 Triangle

Above **Humans "weave" colours harmoniously in many different ways.**

See and hear sound

Jean Dauven, French experimental psychologist, is famous for his research in 1970 proving how light, hence colours and sound, share the same wave patterns. Brain processes used to interpret music are similar to those that turn visual stimulation into images. So close are these processes that in their early days of life babies "see" sound and "hear" colours. Some people have the condition synaesthesia, where they do not lose this ability, continuing to see the colour of sound and music.

Art and architecture

Artists and architects also use colour, composition, shapes, structure and harmony to describe the attributes of their creations. Because music was seen as an expression of universal harmony it was an essential part of an architect's training during the time of Pythagorean influence. This interrelationship of disciplines is evident in medieval times as the Quadrivium, or Fourfold Curriculum, which included mathematics, music, geometry and astronomy. In other cultures, as well, the union of colour and design were, and still are, essential ingredients of sacred places. Acoustics also became an important consideration in European architecture.

Medieval ruins in France

Buddhist shrine

Japanese shrine

English cathedral

Islamic mosque

Above **Gateway to the Heavens, symbolically incorporating energy waves and a colour wheel.**

Moved by the music

In Medieval times students of music regarded themselves as scientists, believing that all forms of harmony were derived from a common source. The meaning of music was more important than its performance. General public interest in musical performance only came about when the Renaissance started to stimulate the development of the arts throughout Europe. From these beginnings we now have a huge range of musical genres and performance styles available to us.

Geometry in space and time

Where there is silence, emptiness and darkness there is no form. From the Centre sound, light and structure move out to the boundary of the eternal Circle, where they bounce back and return to the Point of origin. In the same way ripples expanding outward when a pebble has been dropped are not made by water moving, but by an energy wave moving out in all directions. The water molecules are all held together by gravity. Similarly sound Circles radiate to fashion multi-planed light Beings of matter and spirit.

In the Centre is the seed containing the geometric code that grows and fashions the grids united as the aspects in the Matrix of Space-Time-Being. The original sound, of all sounds possible, carries the code and is split up as individual notes that work together to create music. The original pure white light is split up as the spectrum of colours that combine a wide range of hues and tones. Colours, shapes and sound work together in harmony, composing the music and dance of the Universe.

> *"The distribution of energy follows definite paths which may be studied by means of geometric construction."*
>
> SAMUEL COLMAN, AMERICAN PAINTER,
> INTERIOR DESIGNER AND WRITER (1832–1920)

Sequences, steps and stages of development

Once light splits into the colours and sound into notes, Being can reveal itself so that it can look and learn, grow and become wise through many cycles. Numbers 7, 8 and 9 are steps within the cyclic sequences of development and growth, of accumulated learning turning into wisdom and of the development of consciousness from the sleeping state to Nirvana, like degrees of initiation and steps on the Mystical Path. The 7-fold path appears to be a universal idea in world religions, beliefs and mysticism represented in numerous ways, too many to give adequate coverage here, such as the 7 incisions on the cosmic pole on which the Siberian shaman rises and the walls of the spiritual castle described by the Iraqi Sufi Nuri (c900).

183

The 7 steps lead to the 8th, which is the first step of the next phase; as in the colours and musical scale. Number 7 is the dispenser of life and the source of change due to the passage of time, while 8 is the number of our fate on Earth, the 8 Directions, the Wheel of Law and karmic reincarnation, symbolic of our evolution through many lives and lessons. Number 8 can be halved and halved again: 8 halves to 4, 4 halves to 2, 2 halves to 1 and unity. Because of the even nature of 8 it is the number of justice.

Beyond 8 is 9, the number 3 squared and the number of perfection that lies at the apex of the Triangle. 10 marks a return to the beginning of the cycle, where duality is resolved and the Point is Centred once more.

Above **Assyrian tower (or ziggurat) with 7 levels.**

Right **Giac Lam Pagoda in Ho Chi Minh City, Vietnam.**

Above left and right **7 steps up or down on the soul's path.**

Above **The 7 degrees of initiation into the Mithras cult and the ancient concept of the human soul's ascent into the heavens are the basis of the Christian 7 layers of purgatory as depicted in Dante's images and the 7 steps of the mystical path.**

Women and life

Human life and phases of our development are often divided into 7-year periods. It is of particular relevance in relation to a woman's life and 7-day lunar cycles:
- A girl gets her milk teeth at 7 months and loses them at 7 years
- 2 x 7 = 14 puberty
- 7 x 7 = 49 menopause
- The menstrual cycle is every 7 x 4 days
- Pregnancy is 7 x 40 days from the first day of the last menstruation

Heptagon (or septagon) and heptagram

Number 7 is often regarded as the number of illusion, as the 360-degree Circle cannot be divided by it, as it so easily can by all the other numbers between 1 and 10. Also, unlike the other polygons made out of 3, 5, 6, 8, and 9, 7 cannot be constructed with a compass and straight edge (it requires a marked ruler and compass). The angles in a heptagon total 900°.

Magic number 7

- It is a long-standing tradition that 7 is a magical number, particularly in alchemical processes when formulas were repeated 3 or 7 times and distillations 7 times to be most effective.
- In Islamic magic, the name Abraxas, comprising 7 letters, was frequently used.
- Pythagoreans, notably Nicomachus of Gersa, linked the 7 Planets with the 7 musical notes and 7 Greek vowels.
- 7 is the most important number in the Ismaili Sect (or Severner Shia) and features throughout their texts. Of particular interest is their 7-letter divine word of creation, *kun fayakun*, which is written *knfykun*, meaning "Be!" and "It becomes".
- Along with 3 (the Triangle), 7 is the most important number in the Hindu Vedas and is especially, though not exclusively, linked with Agni, the God of Fire. Agni has 7 wives, mothers, sisters (note the Indian link to the female principle); 7 flames, tongues or beams. Songs devoted to him are 7-fold. 3 and 7 are often combined as 3 x 7 = 21 in the Rig Veda, for example, the 7 rivers grow into 21.

Heptagon

Heptagram

Numbers 1, 2, 3, 4, 5, 6

In the combinations of numbers that make 7 we find 1, 2, 3, 4, 5, 6. Their accompanying attributes are therefore within the power of 7.

1+ 2 + 3 + 4 + 5 + 6 = 21 (2 + 1 = 3) and 21 = 3 x 7
7 is the sum of any 2 opposite sides on a standard 6-sided cubic die.
1 + 6 =7, 2 + 5 = 7, 3 + 4 = 7

A regular (green) and an obtuse (blue) heptagram within a heptagon

German fort design (1604)

Frieze groups

A frieze group is a mathematical concept to classify designs on a two-dimensional surface that is repetitive in one direction, based on the symmetries in the pattern. Such patterns occur frequently in architecture and decorative art.

Left **7 pattern frieze groups.**

Above **Pottery showing the frieze as decoration.**

Indian music – 7 octats
The Western diatonic scale has 7 notes and Indian music has Saptak Swaras, 7 basic notes (sa re ga ma pa dha ni), out of which hundreds of ragas are composed (see page 157).

Below *The Seven Plagues of Egypt* by John Martin (18th century).

Centre **Ancient Biblical 7-headed lion.**

Far right **7-headed dragon or "the Beast" coming out of the sea as seen in apocalyptic visions by John the Apostle in the New Testament.**

Number 7 in Christianity

Number 7 appears frequently in the Bible. Some of the more widely known examples are the 7 days of the week and 7 days of creation.

- 7 Deadly Sins and 7 Virtues
- 7 Gifts of the Holy Spirit
- 7 Pillars in the House of Wisdom
- 7 last sayings of Jesus on the Cross
- 7 churches, 7 stars, 7 seals, 7 last plagues, 7 vials or bowls, 7 thunders in the Revelation, the last book of the Bible

Above **The 7 stellar objects in the solar system visible from Earth – Sun, Moon and the 5 classical planets: Mars, Mercury, Jupiter, Venus and Saturn.**

"The number 7 because of its occult virtues tends to bring all things into being. It is the dispenser of life and is the source of all change, for the moon itself changes its phases every 7 days. This number influences all sublunar things."

HIPPOCRATES, GREEK PHYSICIAN (5TH CENTURY AD)

The chakras

Chakra means "wheel" or "turning" in Sanskrit and it originates from Hindu texts (as far back as the Upanishads – ancient texts considered to be the origin of Hinduism). Typically chakras are drawn to resemble flowers or wheels and are considered the focal points for the reception and transmission of the Life Force.

Typically the major chakras are said to be metaphysical vortices permeating from 7 specific Points on the physical body into the layers of the subtle bodies surrounding our physical body in an ever-increasing cone-shaped formation. They are located along the spine at major branchings of the human nervous system, beginning at the base of the spinal column and moving upward to the top of the skull, through which pass 3 major energy channels. In more recent research, yogic teachings assign specific qualities, such as colour (7 of the light spectrum), Classical Elements (earth, air, water, fire and ether), body sense (sight, touch, taste, hearing and smell) to an endocrine gland. Specific Sanskrit tones, or mantras, are linked with each chakra and used as part of meditation and healing.

Other mystical traditions talk about subtle energies that flow through the body and identify specific parts of the body as having energy Centres, like the chakras, though these vary slightly from one tradition to another in location and number. In Hindu teachings the chakras have a central role as the conduit by which consciousness is awakened, ultimately resulting in union with the divine.

The 7 main chakras (1850)

Octaves and the sequence of growth and development

The Edgar Cayce Readings and the Ra Material (Ancient Egyptian texts) tell us that we are living in a reality based on the octave (7). Numerous spiritual works also say that the One (or Absolute) originated as pure white light that fragmented down the octave, or spectrum, of 7 colours. This light octave is linked to the musical octave in the diatonic scale. Even at a microscopic level cell mitosis (or division) goes through 7 stages. The original seed is the sphere with a Central Point. 7 stages later we have an exact replication and 2 cells: a process akin to the octave-based structure of the atom discovered by English chemist John Newlands in 1865. Once you break through the octave threshold the atom metamorphoses into the next element in the Periodic Table. All living forms, from the micro to the macro, operate in this way, from the sub-microscopic, atomic and cellular levels; up to the planetary scale.

Recapping on Dr Jenny's convictions (see page 140) that our biological evolution was a direct result of vibrations, he believed that each type of cell has its own frequency and that a group of cells with the same frequency in turn creates a new frequency that is in harmony with the original one. Evolution as a process of energetic expansion means that we are going up in stages, up the octaves, up the planes, changing our entire form as we do so, but with the same blueprint.

Shaped by English letters and words

The power of the word can be seen in the structural, descriptive attributes of certain letters and words used in the English language. C words, such as composition and combination, are integral to the binding power of the Circle. All combinations and compositions of colour ultimately combine as one (1, I, Point) and are contained by the eternal Circle. Note the word "one" starts with the Circular letter O. S is the shape of the Spiral Life Force, the momentum that moves energy and makes Being possible through sharing. Shapes provide structure in the fabric of reality and this geometric blueprint is transferred by sound and silence.

> Fall seven times, stand up eight.
>
> JAPANESE PROVERB

GATEWAY TO BECOMING

All art is an imitation of nature.

SENECA, ROMAN DRAMATIST,
PHILOSOPHER AND POLITICIAN (5 BC–AD 65)

Finally, in Part V – Gateway to Becoming – I show how all the elements I have outlined in this book are drawn together to create Nature's rich tapestry and how it pertains to your own experience of life. You will discover more about the holographic Universe and dynamic metaphors of the Classical Elements of fire, air, water, earth and ether. Then you will learn about the extraordinary scope of the human Mind to both analyze and intuit profound levels of understanding. You will gain insights into mandalas and yantras and discover a new symbol that represents the key to understanding reality; a symbol I call the Gateway to Becoming.

21 Holographic Universe

The creation of a thousand forests is in one acorn.

RALPH WALDO EMERSON, AMERICAN WRITER
AND POET (1803–1882)

Above **Past evolution and future forests held in an acorn.**

DENNIS GABOR WAS THE FIRST PERSON to make a hologram (in 1971), for which he received a Nobel Prize. Holography is a type of photography used to record a wave field of light that is scattered by an object. The photographic plate is then placed in a laser, which regenerates the original wave-field pattern as a three-dimensional image. Significantly, every part of a hologram contains all the information possessed by the whole. This theory is linked to the idea that the code for the Universe can be found within its smallest unit, the Point.

Inspired by black-hole thermodynamics, the Holographic Universe theory derives from quantum and string theories and is based on the holographic principle that volume is an illusion and the Universe is really a two-dimensional hologram (or information structure) created out of information etched on the surface of its light-like boundary. Being is revealed by light.

Above **Hologram on a Finnish monetary note.**

Holographic sound

Holographic theory is based on light, the fabric of Being. English acoustics engineer John Stuart Reid proposes that sound is also holographic. Through his research in cymatics (see also page 141) he says that theoretically every atomic particle in a sonic bubble contains all the data of the sound source. When sound propagates in air, every atom or molecule lying in the path of propagation is involved in the process of passing on the sound data that originated from the sound source as smooth repetitive wave motions.

For a tone of single frequency the periodic motion of every atom and molecule will be of the same single periodicity. For complex sounds with a multiplicity of frequencies, the atoms and molecules will each carry this array of periodicities. The human voice is an example of a complex sound and the multitude of

---

Highly efficient information storage

Since 1 cubic cm of holographic film can contain 10 million bits of information, a storage system that is holographic is far more efficient as all information is cross-correlated at an infinite level.

"The Tao that can be expressed is not the eternal Tao;
The name that can be defined is not the unchanging name.
Non-existence is called the antecedent of heaven and earth;
Existence is the mother of all things.
From eternal non-existence, therefore, we serenely observe the mysterious
beginning of the universe;
From eternal existence we clearly see the apparent distinctions.
These two are the same in source and become different when manifested.
This sameness is called profundity.

Infinite profundity is the gate whence comes the beginning of all parts of the
universe."

LAO-TZU,
CHINESE PHILOSOPHER
(6TH CENTURY BC)

"The entire lower world was created in the likeness of the higher world. All
that exists in the higher world appears like an image in this lower world;
yet all this is but One."

MICHEL DE MONTAIGNE,
FRENCH PHILOSOPHER
(1533–1592)

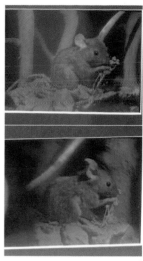

Above **Holographic images of a field mouse, taken from two different angles.**

vibrations describe the uniqueness of each voice. As each atom or molecule bumps into its nearest neighbours their many periodic motions, representing the sonic data of the voice, is passed on at the instant of collision. If we could only see the sound as it is being emitted it would appear as a bubble of sonic energy, the surface of which would shimmer due to every atom and molecule vibrating in unison.

Whole within the part

In the multi-planed Universe the blueprint for the whole remains evident in every part, even without your own body. In 1982, at the University of Paris, a team led by physicist Alain Aspect discovered that under certain circumstances subatomic particles, such as electrons, could instantaneously communicate with each other, regardless of the distance between them. Each particle is aware of what the other is doing. Their evidence supported Bell's Theorem (1964), that subatomic particles are connected beyond three-dimensional Space-Time.

The physicist David Bohm believes these findings imply that perception of an object's reality does not exist and that despite its apparent solidity the Universe is, in fact, like a hologram of immense proportions. Bohm uses the term "holomovement" to convey the idea of a Universe based on a dynamic "web" of relations,

Holographic brains

Neurophysiologist Dr Karl Pribram believes that holographic imagery can be used to explain how memory is stored throughout the brain. Holographic characteristics enable the human brain to store many memories in very little space. He believes that memories are encoded into patterns of nerve impulses that criss-cross the brain in the same way that patterns of light interference criss-cross the plate of a holographic image.

These theories have been extended to include explaining how the brain processes the information from the 5 senses (touch, smell, sight, taste and hearing). Holographic distribution of sensory information is most efficient for translating wave frequencies, for processing sensory input and converting them into our physical perceptions. This has been used to explain how humans can detect the location of a sound without moving the head, even when the person is deaf in one ear.

Above **Symbolically the egg contains the code of life**

out of which all forms of the physical Universe flow. All forms of Being that seem real to us (stars, moons, planets, galaxies, plants, animals) propagate the holographic illusion of our reality. We only see things as separate because we "see" them from the physical plane.

Since every part of a hologram contains all the information possessed by the whole, any pieces of the hologram can therefore be used to reconstruct the whole, rather like a seed. Instead of comprising individual parts to be separated out, information about everything is held within everything, infinitely. If you take something apart in a hologram there are no individual pieces, only smaller pieces of the whole, as in the Mandelbrot Set (see page 67). Separateness is an illusion. This correlates with the idea that contained within the first Point are all the numbers and forms required for propagating all things in the Universe. When the Point sees itself on the boundary of the Circular Void as a passive reflection of itself the principle of duality is established. *1* and *2*, in turn, create the Universe through number and form. The Point is the seed containing the smallest part, containing the whole. Everything is made of *1*s and Is. *1 + 1 + 1*... all the single Eye/I's comprising windows of the all-seeing Eye. By implication, it means that contained within all Beings is the same blueprint of the Universe on all planes and all scales.

> *"This is the truth: as from a fire aflame thousands of sparks come forth, even so from the Creator an infinity of beings have life and to him return again."*
>
> MARCUS TULLIUS CICERO, ROMAN PHILOSOPHER,
> LAWYER AND STATESMAN (106–43 BC)

22 Illumination and intention of the Mind

If you use your mind to study reality, you won't understand either your mind or reality. If you study reality without using your mind, you'll understand both.

BODHIDHARMA, BUDDHIST MONK (5TH CENTURY)

TO BE ABLE TO PERCEIVE MORE REQUIRES AN INCREASE in the scope of our consciousness. The structure of the Triangle (see page 40), with the concept of Being vibrating at different levels, illustrates the point that there are many ways of understanding the scope of perception, or the scope of the Mind. Mind scope is measured by how open and focused the 3 Eyes are (the 2 physical eyes and the Third Eye of intuition), by how liberated the Mind is, such that it may explore its true potential.

Above **A symbolic expression of the expansiveness of the Mind – beyond the limitations of the brain, the head and the physical body.**

"The mind is a miniature Universe and the Universe is the expansion of the mind. He who directly experiences this truth sees the Universe in himself and himself in the entire Universe."

JAKOB BÖHME, GERMAN CHRISTIAN MYSTIC AND THEOLOGIAN (1575–1624)

Mind scope

Mind scope allows us to see, or sense beyond, the physical. The profoundness of this is best illustrated by considering a sensory process. For example, when representing the geography of the Earth on a flat paper map we can represent only shape and form: the undulation of hills, the vastness of deserts and the expanse of oceans. However, we cannot feel the heat, smell the vegetation or be carried by the waves. This additional sensory information expands our knowledge of reality and it allows us to exist within it, with greater awareness.

When the Mind is centred everything coexists in the Moment and we can play with our field of perception. In the "blink of an eye" we have the potential of covering vast distances in space and moving around in time.

"The thought is a deed. Of all deeds she fertilizes the world most."

ÉMILE ZOLA, FRENCH WRITER (1840–1902)

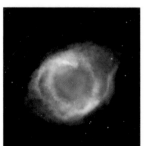

Top and above **"The Eye of God" (The Helix Nebula, NASA)** is a gaseous envelope expelled by a dying star. It is given this name because of its incredible similarity to the human eye.

"The eye of a human being is a microscope, which makes the world seem bigger than it really is."

KAHLIL GIBRAN, LEBANESE-AMERICAN ARTIST, POET AND WRITER (1883–1931)

"It is the mind which creates the world about us, and even though we stand side by side in the same meadow, my eyes will never see what is beheld by yours, my heart will never stir to the emotions with which yours is touched."

GEORGE GISSING, ENGLISH NOVELIST (1857–1903)

Power of intent and directed thought

Marcel Joseph Vogel (1917–1991) was a research scientist who worked for IBM and he received numerous patents for his inventions. His theory was that our thought patterns are vibrations that oscillate like a magnetic field. The Mind and its thoughts are energy patterns in the same geometric fields as matter and directed thoughts act on matter in space. Energy follows thought as light messages that our Mind radiates out.

Symbolically the 2 opposites are brought together in the presence of an independent, neutral third that sits above them and arbitrates, unites, balances and transforms. In the Mind, body, spirit Trinity the third is the Mind. By focusing all 3 of our Eyes thoughts can be directed and this, in turn, directs the energy generated by thought. The prefrontal lobe is significant since it is here that our Eyes are focused and images formed. From our thoughts come words and deeds activated and sustained by both physical and spiritual energy, filled with intent. In this way we can even sense the thought behind an action as simple as a gentle touch. Energy Spirals toward still unity at the apex of the Triangle, concentrating

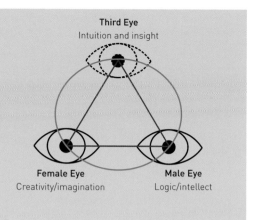

"See" more to perceive more
By opening our Eyes we open up to the true capacity of our Mind. This potential is represented by the Triangle within the Circle that can expand infinitely.

Third Eye
Intuition and insight

Female Eye
Creativity/imagination

Male Eye
Logic/intellect

> Pure essence, and pure matter, and the two joined into one were shot forth without flaw, like three bright arrows from a three-string bow.
>
> DANTE,
> ITALIAN POET
> (c1265–1321)

it in the process, so when the Third Eye is opened and focused this concentrates and focuses the power of insight and precognition. Such power has tremendous impact when directed externally. We can feel this as intent behind thoughts in other people's gazes from across the room. In Hinduism Siva, representative of cosmic consciousness, has 3 Eyes comprising the Sun, Moon and fire. Siva's third fire Eye allows him to destroy whatever he looks at.

Our thought precedes our speech and our action. The energy of each and every thought we have is carried through to what we say and what we do. Within our thought, and hence words and deeds, is the power of our intent. And this power originates from our Mind.

Out of sight, out of mind

Light induces a chemical change in photosensitive chemicals on photographic film. Human sight is similarly based on chemical changes in the eye, which generates nerve impulses that our brain interprets as colour, shape and location of objects. Light creates our reality; what we see. To others, we are just images of light received by their eyes; we are holographic light images projected into the physical plane, like a movie of cosmic proportions. Hence holographic reality may be viewed as an illusion of light images woven together.

Reality is a symphony of sensory images, of surface features. Our physical senses are limited and constrained by our material forms (in 3D space and time). Our infinite Mind is trapped in a physical body, aptly conveyed by the symbols of an Eye in the middle of a Triangle. What we perceive to be complete is but a minute piece of the complete picture; most of it is "invisible" and outside our physical senses.

Holding a candle to the soul

Many people with scientific backgrounds are now seeing the links between the light of spirit that features extensively in mystic and religious texts and the physical light spectrum. Through the windows of our physical eyes enter light images

Above **Robert Fludd's image of the Mind incorporates the physical senses, visual impressions and a link to the spiritual Planes.**

Above **Fibre optics carrying data as light.**

of the physical world comprising forms, or crystallization of energy as forms, shaped by patterns. The Mind is illuminated by the knowledge supplied by light entering our 3 Eyes and this light guides the soul toward enlightenment. Candle flames are a symbol of an individual soul and the flame that is the presence of "heavenly light" and the radiance shed by wisdom and truth that lightens the darkness of ignorance.

The pagan mystics taught that each individual is a spark of the one fire of life, as represented by the Sun. Buddhists liken the Higher Self to a flame burning on a candle of the personal self. When the flame completely consumes the candle, burning it totally into light, both flame and candle no longer exist, so symbolically there is neither personality nor soul. The individual flame returns to the one source as light. The realization of this truth is "enlightenment" and the state of Oneness is called "Nirvana", meaning "to extinguish".

No wonder the Sun has been revered and placed at the Centre of the Universe. It is the primary source of light, energy and life. Phrases such as these and "energy entering the Eyes to reach the Centre of the Mind", "acting as a messenger that affects life", "no rigid structure to matter" are clues that light may be consciousness, the very basis of our Being and the fabric of life and form.

Illusion, order and intent in Being

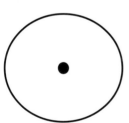

Top **This photograph encapsulates the Sun glyph and even includes the Duals of active and reflected light.**

Above **The Egyptian, Mayan and Chinese glyph for light, also known as the Sun symbol.**

Underlying every Plane of Being are the same patterns, the same order. In a holographic Universe any fragmentation is artificial. As Plato suggested, on the lower levels of vibration of Being we see "shadows", or reflections, of the true totality of reality. A Universe made of holographic fields establishes a framework that can be used to explain many occurrences, such as the power of intent, how we can affect each other over distances without being immediately next to each other. It would be possible to explain how the extended senses, such as telepathy, would work – for example, sensing who is on the other end of the phone when it rings. If the waves making up the planes of the Universe are holographic then any part of the Universe is accessible through the use of our Mind.

Our Universe is a beautiful, artfully constructed place, resonating harmoniously. It is fine-tuned, full of subtle cause-and-effects and incredible coincidences that cannot possibly be random. Since physical reality is a reflection of the higher Planes and Dimensions of Being, just as two-dimensional maps are simple reflections and representations of three-dimensional geography, it is by going "up" that we can come back "down". The Matrix of Space-Time-Being that binds the many into One extends from the infinitely small (microcosm) to the infinitely large (macrocosm) and beyond into other dimensions of reality.

"In the supreme golden chamber is Brahman, invisible and pure. He is the radiant light of all lights."

THE UPANISHADS (PRE-BUDDHIST)

23 Gateway to the Heavens

In the final analysis, a drawing simply is no longer a drawing, no matter how self-sufficient its execution may be. It is a symbol, and the more profoundly the imaginary lines of projection meet higher dimensions, the better.

PAUL KLEE, SWISS-GERMAN ARTIST (1879–1940)

REALITY CONSISTS OF A GRADED HIERARCHY OF CONSCIOUSNESS, which corresponds to the various vibrating planes of Being. Higher-Plane information, which is finer in vibration and intangible, is dressed up as visual images so that it is comprehensible for our physical brain. On the physical Plane, it is not possible to occupy the entire Mind's scope, only portions of it. Our physical brain is bounded by tools in their own right, so although we can contemplate the infinite and the fact that it exists, we can distil our Mind's thoughts only physically and so our descriptions of the infinite are necessarily physically finite.

Through imagery the timeless, formless and boundless are made concrete using symbols. Though the Gateway to the Heavens model may appear basic and superficial on the surface, in essence it is a highly complex model of the Universe. To see more we need to look beyond its surface and think of it as a thought-form. We need to observe it from each Point of view, investigate it using logic, explore with the imagination and sense it intuitively. Then from the middle we should blend all 3 perspectives and let in the light of knowledge to learn and experience reality more intensely, by expanding the Mind's scope. It is possible to reach the stage (and Point) where the individual, symbols and the Universe are One.

> Enhance and intensify one's vision of that synthesis of truth and beauty which is the highest and deepest reality.
>
> OVID,
> ROMAN POET
> (C43 BC–AD C18)

Original holographic model

The Gateway to the Heavens is a universal geometric model found across cultures, disciplines and belief systems. It begins and ends with the Centre, the Point, from which everything originates and returns and around which everything revolves. Duals provide the interplay in all the contradictions and contrasts, pulling and pushing life, testing our decisions. From this Point (number 1) and the Duals (number 2) a web of reality can be spun as the impenetrable Matrix of Space-Time-Being. The Duals' effects are within each of the shapes and help define their boundaries, reminding us of the apparent limits of our perceptions. The Circle and Square define where and when we are in time and space. Triangles pertain to the framework manifesting energetic Beings. And so it is that the single

Point, Duals, Circle, Square and Triangle effectively construct a Matrix, which is the stage and upon which which life is acted out.

In the process of the construction of the Matrix other principles emerge that are fundamental to the animation and purpose of reality. Spiralling cycles of the Life Force animate and generate the myriad forms of Nature and enable us to grow and evolve as one with the Universe. Inspiration and guidance are provided by the tests of the Cross, which highlight the Central Point. The Cross reminds us why we are here, where we are going and what we will eventually become. Together the Cross and the Spiral direct the play, provide the storyline and the underlying purpose to reality.

> **Catalyzed by letters used in the English language**
> My own personal catalyst for writing came about by seeing letters of the English language turn into geometric shapes in my imagination. These letters, for example C and S, bear a resemblance to the Circle and half Circle. Words that begin with these letters describe concepts that also relate to the Circle – for example "community" and "continual".

Yantra and mandala

"Yantra" and "mandala" are particularly associated with Hindu and Buddhist traditions and practices respectively, but mandalas, in particular, cross over these belief systems. In addition, the words yantra and mandala are used interchangeably, but there is a subtle distinction between them. "Mandala" has become a generic term for any plan, chart or geometric pattern that represents the cosmos metaphysically, or symbolically, as a microcosm of the Universe. Even though mandala means "circle" in Sanskrit, in its most basic form it comprises a Square with 4 gates in the middle of each side that face the 4 cardinal Points. This Square contains a Circle representing the concentric Circles radiating from the Centre as a Holon, surrounding a central Point, as in the Gateway to the Heavens model. Both Buddhist and Hindu sacred art and architecture frequently take the form of a mandala, which may then be used for establishing a sacred space, to focus attention, as a spiritual teaching tool and as an aid to meditation.

Think of yantras as spiritual technology working on multiple dimensions. They are tools, or thought-forms, for holding and conducting energy patterns (often for their mystical or magical powers). This concept derives from the root words of yantra that mean "to hold and sustain". Not surprisingly, the mystic yantras are a combination of 3 principles – form, function and power. The form therefore has a function and holds power or energy.

Top and above **Examples of Buddhist mandalas.**

Each yantra is unique, employing many geometric shapes and patterns in various combinations – including Squares, Triangles, Circles, floral patterns and symbols such as the swastika. So a mandala is a type of yantra that embodies the energetic purposes of the basic shapes. Yantra Yoga (also known as Mandala

Second Stage to building up the Gateway to Becoming

The dynamic shapes and symbolism of the Cross and Spiral are added to the structural Matrix of Space-Time-Being. Incorporated within the static Matrix it is activated into the Gateway to the Heavens model. Though the Cross is evident, the Spiral is not. It is a subtle force animating every feature.

KEY QUESTIONS AND SHAPES

Question	Shape	Centre	Active word	Aspect of reality
When?	Circle	Moment	Now	Time
Where?	Square	Point	Here	Space
Who?	Triangle	Eye	I am	Being
How?	Spiral	Sharing	Because	Life force
Why?	Cross	Choice	To	Centre

SUMMARY MEANINGS OF THE KEY SHAPES
Their associated descriptive words and first letters

	Circle (C)	Square	Triangle	Cross (C)	Spiral (S)
Prime letters	C	D	V	I	S
Structural words	Communicate Commune Connect Community Continual Change Cycle	Domain Distance Direction Domicile	Vibration Vessel Vision/View Variety Value Void	I Interconnected Individual Chance Choices Cross	Share Sustain Symbiotic Stability Survival Symphony
Active words	Evolution Events Experience	Movement Motion	Energy Ethereal Existence	Surrender Search Strength	Harmony
Measures	Second, eon	Line, area, volume	Sentience, vibration		

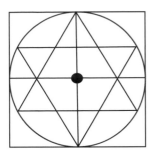

Gateway to the Heavens

The Sri Yantra

Above **The Sri Yantra outline, embossed on copper.**

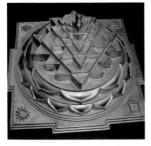

Above **The Sri Yantra shown in three dimensions.**

Yoga) is the path of union with the Absolute through geometric visualization, combining the external and internal experiences of reality as one.

The Sri Yantra

Numerous examples of geometric models similar to the Gateway to the Heavens model exist, but the most similar is the Sri Yantra (see left), which also contains the essential features of a mandala. Sri Yantra is translated as "gateway to the heavens". Imagery in the Sri Yantra is primordial and there are many references to it in the Hindu texts, supporting its importance as a universal symbol. The Sri Yantra is read as a chart, from creation through the stages of evolution of the Universal. Within it are each of our own evolutionary paths and life purposes, through many lifetimes, spun together and enclosed by the Circle of infinite time. Each Sri Yantra is either constructed using the order of "evolution" so that the yantra is created from the Central Point (bindu) outward, or the order of decomposition from the Circle inward. Once completed, the yantra is consecrated and given sacred power through a ceremony called "the opening of the eye".

Adding the dynamics of the Classical Elements

Not so long ago our ancestors lived intimately within their environment and its spirits and viewed themselves as part of Nature, observing the creative process within. Earth was sacred, not a material thing made of rock, but a spiritual Being, a living entity. Other created Beings were honoured, not taken for granted. They understood their relationship and utter dependence for survival on connections, cycles of life and the interplay of opposite forces in their environment.

While geometry is structural, the Classical Elements are dynamics within Nature that add another layer of knowledge about the abstract forces that exist within and beyond the physical. Each Element has a unique purpose and role, just like each of the basic geometric shapes. However, their roles pertain to the creative process and nature of Beings. By adding the Classical Elements to the Gateway to the Heavens model we enhance the power of 5 beyond the Spiral, revealing further its fundamental role in the facilitation of creation. The dynamics of Being and the personality of created Beings becomes evident.

Human microcosm
Our human body is structured by geometry and it contains the different Elements, so it is a microcosm of the Classical Elements in action throughout Nature. We are comprised of ether, in the form of the Life Force, air as in the oxygen we breathe, liquid in the forms of various biochemicals and also a large percentage of water, solids in the form of food and the minerals and carbon, which comprise our various cells. Heat within us transforms chemicals for our use – and it keeps us warm.

24 Earth

Each blade of grass has its spot on earth whence it draws its life, its strength; and so is man rooted to the land from which he draws his faith together with his life.

JOSEPH CONRAD, ENGLISH NOVELIST
(1857–1924)

THE ELEMENT OF EARTH IS LINKED WITH THE STATE of matter known as "solid"; not to be confused with Planet Earth. Earth element is the solid, bodily aspect of Being and so is considered to be the heaviest of the Classical Elements. Rocks and crystals are the oldest material forms of Being and they symbolically and literally provide a firm foundation and platform, or stage, upon which life can act.

On this stage physical Beings flourish. In vast quantities lush plants provide a multitude of animals of all shapes and sizes with sustenance, habitat, building materials, protection and even warmth. When plants and animals die their physical body is recycled, so in essence we also potentially become part of a rock, a tree or another animal. Our bodies are manufactured out of minerals, plants and other animals using the same blueprint, the same rules and principles and animated by the same Life Force. There are no physical boundaries in the earth element, only changes in physical form.

Above **Salt's cubic structure epitomizes the essence of the 4 Square, solid foundation of the Earth element. It is an essential mineral in our diet for our physical survival. Since salt purifies and cleanses, it is often used in rites to purify a sacred site.**

"There is a road in the hearts of all of us, hidden and seldom travelled,
which leads to an unknown, secret place.
The old people came literally to love the soil,
and they sat or reclined on the ground with a feeling of
being close to a mothering power.
Their tepees were built upon the earth
and their altars were made of earth.
The soul was soothing, strengthening, cleansing and healing.
That is why the old Indian still sits upon the earth instead of
propping himself up and away from its life giving forces.
For him, to sit or lie upon the ground is to be able to think more deeply
and to feel more keenly. He can see more clearly into the mysteries of
life and come closer in kinship to other lives about him."

CHIEF LUTHER STANDING BEAR,
NATIVE AMERICAN WRITER AND ACTOR (1868–1939)

Above **Conical Native American teepees resting upon the earth.**

Summary diagram of interdependent earth cycles

Each circuit is dependent upon the inner circuits to sustain its existence.

Inner circuit – **rocks and minerals continually change state and form due to pressure, heat and erosion.**

Middle circuit – **earth provides a bed and nutrients for plants to grow and build themselves. In turn they die, return to the earth and nourish it.**

Outer circuit – **earth provides a home, shelter and nutrients for animals (for example salt) and in water, through consumption of plants and, for some, by eating other animals. Through excrement and in death, animals also return to earth.**

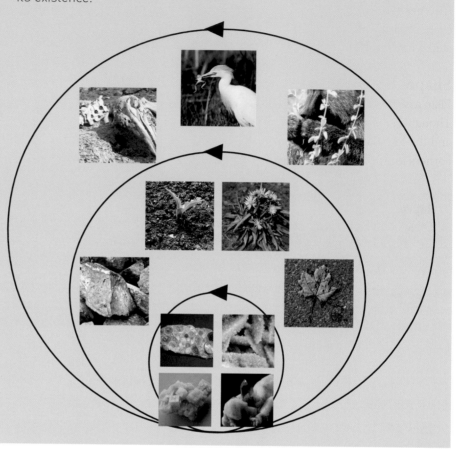

> What springs from earth dissolves to earth again, and heaven-born things fly to their native seat.
>
> MARCUS AURELIUS, ROMAN EMPEROR (AD 121–180)

Caves and mountains

While mountains are places associated with transcendent sight, caves are associated with insights gained from going into the dark within. Caves and crevices are primal earth wombs, the first places man used to intimately link with the Earth Mother.

Feminine element earth

Earth is solid and enduring and is the ultimate female element. The Planet Earth itself is a sphere of life and as the year turns, we can watch life being nurtured as it comes from, and returns to, the soil, to be used and reborn perpetually in the cycle of birth, life, death and rebirth. For this reason the Planet Earth is associated with Goddess Energy.

Shapes of the earth element

Plato assigned the cube to the element of earth. In 2 dimensions this is the Square. Though Planet Earth is a sphere, when we are on her surface we see the Square and the foundations of our existence. This is why the ceremonial laying out the 4 corners for the foundations on the Earth is so important in many traditions.

Representations of the earth Element

These are the main symbolic representations of the Element earth. It is interesting how the astrological symbol for Planet Earth reveals significant links to symbols for the earth Element.

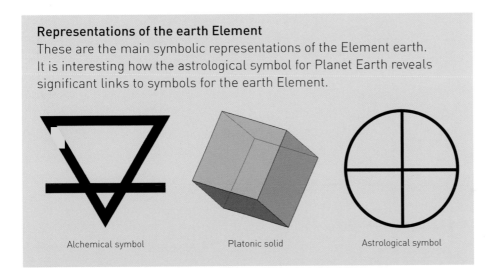

Alchemical symbol   Platonic solid   Astrological symbol

Colours of the earth Element

Earth looks brown and is formed by the combination of colours derived from plants and stones, but symbolically it is represented by the slowest, densest, most physical and grounding colour, red. Scarlet is the colour of our blood, which ties us to the Earth.

> Life comes from the earth and life returns to the earth.
>
> Zhuangzi,
> Chinese philosopher
> (4th century bc)

25 Water

> In the world there is nothing more submissive
> and weak than water. Yet for attacking that which
> is hard and strong nothing can surpass it.
>
> LAO-TZU, CHINESE PHILOSPHER
(6TH CENTURY BC)

Water on the Earth's surface

THE ELEMENT WATER IS THE STATE OF MATTER known as "liquid". Water, H_2O, is the elixir of life that flows through the body and gives it form. Perpetually circulating over, under and through the surface of the Earth and all the forms of life on her surface, water acts as a carrier, recycles and cleanses. Moving above and below the surface of the Earth as oceans, rivers and springs; through the air as vapour, clouds and rain, the water in our own body has existed in all these locations, and eventually will return to them.

Life-bestowing water

The stuff of myth, water has been linked to the mysteries of human existence for as long as history has been recorded. For this reason many legends are based on water-bestowing "life", sentient consciousness and knowledge of existence, wisdom and immortality. Worldwide, clear springs of water are welcomed and frequently honoured for their healing qualities – whole cities have been built up around healing spas. Fountains are frequently important symbols of immortality and in Buddhist temples dragons, linked to water through myth, are often used as fountainheads for purification of water before worship.

Water can be "charged" by movement, Sunlight and the earth (as in pure spring water), hence the calming yet energizing effect of being near running water, especially a powerful waterfall or crashing ocean waves, which create a negative-ion-rich air.

Water under ground

Purity and spirit

Water needs to move; if it stays still it stagnates. It will not be contained, so it takes its course and flows on, regardless of barriers, in a sense reminding us how important it is to embrace the natural flow of life. The spirit of water is purity. Symbolically baptism with water cleanses the individual from sin and marks a symbolic rebirth. Immersion is a return to liquid potential, before you are formed, so that you can come back regenerated with your Life Force intensi-fied. On a larger scale, this is the case with floods, dissolving all forms so that the

Water in the Earth's atmosphere

States of water and the Grid of Being

A unique attribute of water linked to the Grid of Life is the "triple point" (see page 109). This is the temperature at which ice (solid), water (liquid) and vapour (gas) co-exist under a very specific level of pressure.

When water freezes each molecule bonds with 4 other molecules and forms crystals that collect to form pyramidal shapes that are hollow and filled with air, which is why ice is lighter per unit of volume than its liquid state. Unlike other chemicals water expands when frozen and then floats on its remaining liquid, rather than sinking.

As the temperature drops water molecules vibrate more slowly and electric charges within each molecule attract other molecules, to form a tight-knit hexagonal arrangement. 6 water molecules are at the core of every snowflake. More molecules build up on the core of 6 to become unique snowflakes. Though each one is unique they are all restricted to only 1 pattern, reflected and repeated 12 times within a hexagon. In simple snowflakes we see the infinite variety of life facilitated by the Grid of Life.

Top right **Ice and water at Niagara Falls.**

Centre **Water and steam at Old Faithful geyser, Yellowstone National Park, USA.**

Right **Hexagonal ice crystal**

Creation and destruction

Fine water in the Earth's atmosphere filters the Sun's rays to make life possible on Earth. Science also links water to the source of life. Indeed most creation myths cite water as the element that gave rise to all life, usually as a cosmic ocean. Water may be the source of our origins, but it also represents destination. Certainly water is also the most powerful destructive force. Over 500 myths tell of the destruction of the Earth by water, heralding the dawn of a New Age.

Above **A Shinto shrine which uses spring water for a purification ritual.**

Left **The power of water erodes rocks spectacularly and in floods brings life as well as destruction.**

Top and above **Pure moving, living water and a stagnant pond of polluted water.**

Above **This flow form at Chalice Well, Glastonbury, England, is a waterfall featuring 7 bowls shaped like parts of the human female reproduction system. Water flows in a figure-of-8 movement before entering the vesica piscis-shaped pool via a male phallic symbol.**

elements can be recombined in new patterns. Clear, clean water cleanses, but it is also very susceptible to pollution, which has far-reaching destructive effects on the fabric of Earth's life. Stagnant water is still and opaque and though it may be used to quench thirst only crystal-clear, moving, living water is believed to heal body and soul because its own spirit attributes are vibrant.

Water – the feminine element

Water is a feminine energy and like earth is highly connected to aspects of the Goddess. It is less dynamic than air, but less static than earth. Some 80 per cent plus of the human body comprises of water, supporting the myth that we are made of clay; the feminine elements of earth and water.

Watery emotions and feminine intuition

In Ancient Greek philosophy and science, water was commonly associated with the qualities of emotion and intuition (dreams, divination, intuition and psychic abilities). There is a direct link here with the effect of the Moon on water, specifically the ocean's tides, currents and the ebb and flow of emotions. Our emotions may be calm and gentle or violent and destructive. Indeed, when we are emotional we shed tears or sweat. Both these bodily fluids comprise saline solution made of two key life-sustaining elements: water and crystal salt. The power of emotion, like water and salt, is also a medium for purification and healing. The Moon is also associated with intuitive abilities, cycles in fertility and gestation. In Hinduism the element water is also associated with Chandra (the Moon) and Shukra (Venus), who represent feelings, intuition and imagination.

Shapes of the water element

The Platonic solid associated with water is the icosahedron. Formed from 20 equilateral Triangles this makes water the element whose Platonic solid has the greatest number of sides. In an aerial, two-dimensional, view it is pentagonal.

Water moves relentlessly. As a driving force it is powerful as it flows and runs its course. Frequently Spirals are used to denote the life-giving properties of water and its Spiralling motions.

The spirit of water

A team led by Masura Imoto, a doctor of alternative medicine and entrepreneur, has studied the effect of external stimuli, such as thoughts, music, sound and words, on ice crystals. His images reveal how water crystals are dramatically affected by these stimuli, even intent in directed thoughts. Negative sources and crystals are distorted; positive sources result in beautiful, perfect crystals. As our bodies are primarily water, so the implications of the quality of the water we use on our own health and well-being and also external stimuli on our internal waters, are obvious.

Water and the Moon

Peter Filcher, in *God's Secret Formula*, highlights an interesting attribute of water that links with its female aspect. The period of growth in the human womb between conception and birth is 10 sidereal months (each month takes 27.32 days) or 273 days. The female oestral cycle follows the true astronomical rhythm, not full Moon to full Moon, which is two days longer. This figure of 273 days is the same number as absolute zero of −273 degrees C, a fact which amazed Filcher and reinforced his view that water and the Moon are connected in some way beyond the obvious influence the Moon has on moving water over the Earth's surface. He highlights the fact that mysteriously every 55 millionth water molecule is split such that water is always an inseparable mixture of *3* components; H_3O+, H_2O and $OH-$.

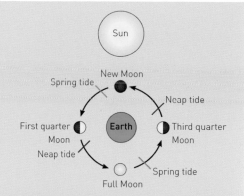

Above **Schematic showing the influence of the Moon on tides. Neap and spring tides reach their maximum force 2 days after the first and third quarters, new and full Moon. Spring tides happen at full and new Moons and neap tides at the one-quarter and three-quarter Moon. Every 6 hours the water also lowers or rises, making 4 tides.**

Water as a mirror

Water reflects images, just as the Moon reflects the Sun. Like a mirror reflecting the solid, physical world as an ethereal image, each world appears the same though it is made of a different substance. At another level, since water is reflective, by looking at its surface you see your reflection and explore the depths of your Being to find your true Self.

Far left **Alchemical symbol for water.**

Centre **Platonic solid.**

Left **Spiral symbols on water jars**

> Water is the driving force of all nature.
>
> LEONARDO DA VINCI,
> ITALIAN RENAISSANCE
> POLYMATH
> (1452–1519)

Colours of the water element

Pure water is crystal clear and colourless, though it is most frequently depicted in various tones of cool blue.

26 Fire

> The fire which enlightens is the same fire which consumes.
>
> HENRI-FRÉDÉRIC AMIEL, SWISS PHILOSOPHER, POET AND CRITIC (1821–1881)

> Heat cannot be separated from fire, or beauty from The Eternal.
>
> DANTE, ITALIAN POET (c1265–1321)

TEMPERATURE IS A CONCEPT THAT IS DIFFICULT to explain in physical terms, but essentially it relates to movement or changes in position in Space-Time. Temperature is not a dimension, but an inherent part of the state of Being, the nature of which is still relatively unknown. Like water and the nutrients of the soil, a warm temperature is necessary for physical life to exist. Temperature variations change form and are also a by-product of change.

Extreme cold, or the lack of the fire element, stagnates life and extreme heat destroys it. The Ancient Greeks distinguished the destructive and consumptive (*aidelon*) heat of fire, associated with Hades, from the creative fire associated with Hephaistos, the God of Metalworking. Balance and stability is essential so that heat at the right temperature in a stable environment stimulates creation and the chemical processes necessary to procreate Earth's life. Also the narrow range of temperature in which water is liquid, the elixir essential to life on Earth, may be the optimal one for life to exist spontaneously. Earth, water and heat (fire) work together seamlessly and subtly in the process of creation.

Hexagonal heat convection and the Grid of Being

In 1901 scientist Henri Bénard noticed that when liquid is heated uniformly a constant heat flux moves from the bottom to the top via conduction. The liquid stays still in a state of equilibrium and the temperature is the same throughout. Eventually a point of instability is reached, when temperature difference between top and bottom reaches a critical level and heat convection comes into play. Ordered hexagonal cells are spontaneously formed. The hot liquid rises through the centre of the hexagonal cells and the cool descends to the bottom. This form maintains the non-equilibrium state by allowing the heat to flow throughout its hexagonal structure. This is known as the Bénard Instability, regarded as a classical case of self-organization. In Nature the same process occurs in a variety of conditions. For example, the flow of warm air from the Earth's surface toward outer space generates hexagonal circulation vortices, leaving images on desert sand.

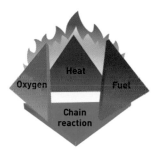

Above **Fire Trinity: the 3 parts needed for fire – heat, fuel and oxygen.**

Heat: the source of order and pattern-forming

Studies on spontaneous pattern-forming due to the application of heat have resulted in the view that heat is a source of order, not a wasteful by-product. Thermal studies and research on cyclical feedback loops show us how self-sustaining systems using patterns of order become more and more stable as their complexity increases. Our bodies are examples of complex self-sustaining systems in which energy and matter flow continually through the system in stable states over long periods; states that are not even close to equilibrium. Because they are stable, though complex, the same overall structure is maintained even when there are changes in its components, like removing a body part. Life is made of these incredible "living" self-sustaining, pattern-forming systems that are effectively ordered by heat. The most obvious example is weather patterns.

Above **Within the "male" there is an innate attraction to fire.**

Fascination with fire

Our ability to use fire opened up a unique path in human evolution. Early humans gained control of fire around 1.4 million years ago. Since then we have used fire for protection, for light, for clearing land, for cooking, in our experimentation and knowledge-gathering and within our spiritual traditions. Fire has played an important part in all cultures and continues to do so in modern times.

Left **Fire in action: lightning, volcanic lava, the Sun and forest fires.**

> Zeal without knowledge is fire without light.
>
> THOMAS FULLER,
> ENGLISH CHURCHMAN
> AND HISTORIAN
> (1608–1661)

Agent of physical transformation

Energy cannot be created or destroyed, but it has the ability to transform endlessly; to change. Before the atomic theory of matter, melting, burning and freezing were regarded as distinct and different phenomena rather than the characteristics of just one process. The 3 forms of matter – solid, liquid and gas – are organized collections of atoms held together by bonds that can only be broken by temperature or pressure. Introducing temperature changes, or pressure, alters the bonds holding physical matter together and changes the state in the material to produce a new form, often irreversibly, just like cooking. Using water (see page 205) as an example, we have seen the changes of state in one process clearly from a number of frozen forms (such as ice, snow, sleet) through to liquid and vapour.

Fire and spiritual transformation

Having the power of transformation of Being, fire transcends and consumes form, releasing the old to embrace the new. Because of this it is associated with the spark of life and reincarnation and the ability for the life to return to a body once more. Many religions associate fire with light and spirit. A common practice of ancient cultures, such as the Egyptians, Greeks, Incas, Persians and Romans, was to keep a sacred fire burning in the centre of the village. The Cherokee Native Americans kept the Most Sacred Fire burning continually within a large 7-sided building, to hold the spirit of the tribe together. Similarly, in many households, the fire in the hearth is a focal place where families and groups gather for warmth and also a sense of community and protection.

In belief systems fire plays a major role in spiritual rites and passages of initiation. Alchemists believed that fire was the agent of transformation and that all

Top and above **Single flames that originate from the original flame: Buddhist monks lighting candles.**

Flames

Intensely hot flames are ethereal blue and clear, while cooler flames are redder and more "solid" in appearance. Appropriately, higher temperatures reduce matter to less and less "solid" forms, from solid to liquid and then to gaseous states. More solid forms are cold and the more ethereal are hot.

The masculine element fire

Fire is "alive" and its primal energy has always fascinated humans, particularly men. It is not surprising that fire is a male element. Having multiple roles, fire purifies, is linked to the flame of will and energy, creates and destroys, heals and harms. Fire also symbolizes the fertility of God as the many sparks of fire, or light, that comes from the original flame radiating from the Centre.

things come from, and return to, fire. As outlined on page 196 it is a common metaphor used in belief systems that each individual is a flame that will eventually return to the One as pure white light when the ego has been consumed by fire. Realization of this is known as "enlightenment".

Light the flame within

Fire also represents the creativity and passion that all intellectual and emotional Beings have. The Element is also very quick to "flare" or "ignite". Through fire we can connect with this inner flame, often represented as a pyramid, to revitalize our zest for life.

Shapes of the fire Element

Plato associated the tetrahedron with fire comprising 4 Triangles. Tetrahedrons contain the least volume of any Platonic solid with the greatest surface area.

Alchemical symbol for fire

Platonic solid

Salamanders: "I nourish and I extinguish"

The legendary salamander is a typically lizard-like form, but it is usually seen as having an affinity with fire (sometimes specifically elemental fire). This is probably because many actual species of salamander hibernate in and under rotting logs. When this wood was put on the fire the salamanders mysteriously appeared from the flames. The salamander became a symbol of enduring faith which triumphs over the fires of passion and was the badge of Francis I of France, with the motto, "I nourish (the good) and I extinguish (the bad)."

Salamander emblem

Phoenix and rebirth

The phoenix is one of the most well-known symbols of purification and transformation. The phoenix has become a universal emblem of death by fire, and of light, resurrection and immortality. The firebird sets itself alight every hundred years and rises from the ashes, exalted. In Mexico, the great God Quetzalcoatl was accompanied by the phoenix, while in China the bird represents the Empress and when it is with the dragon (the Emperor) it stands for inseparable fellowship. To early Christians the phoenix was a representation of Christ.

Above **Phoenix emblem (1697); "It rises again more glorious."**

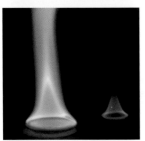

Top and above **Orange is primarily the colour of fire, though, by association, so are the warm colours of yellow and red. Blue flames, being extremely hot, are more ethereal.**

Fire temples
Ancient Vedic fire altars, *Pranga-cit*, were constructed from simple geometric shapes that were of celestial significance. Altars constructed in the form of the "magical triangle" were associated with rites for gaining strength and supremacy. These followed exactly the same format as the Grid of Being.

Far right **A Srividya *homa* (fire sacrifice) to the Goddess, with the hearth in the shape of an inverted Triangle.**

Right **The geometric form of a "magical triangle".**

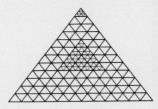

Eggs and the Sun
The element of fire appears in mythological stories all across the world, often in those related to the Sun. Frequently the Sun is symbolized as a Circle with a Central Point representing the "I am" that is in the process of becoming. With light (the Sun) came the rainbow of prism colours as it shines on and through water. This analogy is literally reflected in the "egg", which has a Sun-like yellow orb centre surrounded and supported by salt water. Saline water surrounds the fertilized nucleus, protecting it and keeping it sterile, like the water in the womb.

Sacred fires, traditional and modern
A sacred gift of the gods, fire is used to see into the opposite realm of the spirit. Native tribes dance around fires in order to call forth spirits and monks chant while gazing into a single flame. Modern sacred fires include examples such as the Olympic flame.

Eternal Flame, Arlington Cemetery, USA

A ritual marking entry into Hinduism in Nepal

Fire-making ceremony, Borneo

Olympic flame, Vancouver, Canada

27 Air and ether

All things share the same breath - the beast, the tree, the man...the air shares its spirit with all the life it supports.

CHIEF SEATTLE (c1780–1866)

AIR IS A PURE SUBSTANCE THAT HAS NO FORM and has a power to animate Being. Element air has the state of matter known as "gas". Air comprises not only oxygen, but also all other gases, such as carbon dioxide that plants require for photosynthesis, which in turn produces oxygen. In the beginning of the Universe only the ingredients within gas existed. These combined to create matter as the Universe matured. So gas is also latent Being.

Seeing the effect of air

Changes in pressure and temperature move air, otherwise it would be immobile. Like water, air moves in Spirals and eddies. But, having no physical form its motion can only be seen through the effects it has on solids and within solids. This is why smoke and incense are frequently used to represent air. Also, like water, air circulates. It is inhaled and exhaled to blend with the atmosphere, passing through physical forms of Being in continual cycles.

> Wind is a floating wave of air, whose undulation continually varies.
>
> MARCUS VITRUVIUS POLLIO, ROMAN WRITER, ARCHITECT AND ENGINEER (c80/70–c15 BC)

Air and water
Characteristic movements and patterns of water and air are used in many cultures as metaphors to illustrate the Dual characteristics of energy as matter and spirit respectively.

Alchemical symbol for air

Platonic solid

Alchemical symbol for
ether/spirit

Platonic solid – icosahedron for
ether

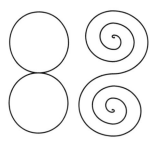

Infinity symbols for the Life
Force, physical and spiritual air
sustaining matter and spirit

Sound carried by air

If air remained still there would be no sounds carried to our ears. Air also creates sound due to its impact on physical forms, such as rustling leaves. Hearing the motions of air informs us of the motion of sounding, or vibrating, forms of Being. Since air carries sound it also helps move the geometric code contained in the sound's vibrations.

Breath of life – elements of air and ether

Element air animates and sustains Being on two levels. The Ancient Greeks used two words for air: *aer* meant the denser lower atmosphere, and *(a)ether* meant the bright upper atmosphere above the clouds. These effectively correlate with the air of the Earth's atmosphere and the fifth Classical Element of ether, prana, *chi* or the Life Force. Terrestrial air containing oxygen, represented as an O (a Circle), physically fuels us and also the element fire. Spiritual breath, or prana, fuels our spiritual aspect and is of the Heavens "above" Earth. Each type of air is extracted through methods of inhalation to fuel our spirit and blood, keeping it a rich red colour. Without air we would perish immediately.

Let the imagination take flight

Air has no form; it is limitless and unbounded like wisdom – the potential powers of the Mind and imagination. This is why air is linked to creativity, wisdom and intelligence, contemplation and study. Like air the imagination is free to fly.

Fresh air oxygenates our blood and brain to clear the head. The more ions in the air the better it is for our health (ion-rich air is found in mountains, by the ocean and in pine forests).

Prana and Chi

In the Hindu tradition Prana is a primary deity and the father of Bhima, the spiritual father of Lord Hanuman. As the word for air or wind, it is one of the "5 great elements" in Hinduism. The Sanskrit word *vata* literally means "blown" and prana means "breathing" (the breath of life). The technique for inhaling prana is a major part of yogic teachings and meditation. Stressing the importance of conscious breathing it is a technique that allows us to more effectively absorb the Life Force that surrounds and permeates us at all the Planes of our Being. Breathing consciously in meditation heightens awareness.

The Ancient Chinese concept of *qi*, or *chi*, corresponds to the concept of prana. *Qi* or *ki* (Japanese) is believed to be a Life Force that is part of every living thing (see page 49). It is often translated as "energy flow", or literally as "air" or "breath". For example, the Chinese word for weather means "sky breath".

Left to right **A heavenly host** – an angelic choir; the symbolism of wings of angels and a dove; bronze angel statue, Canada. Angels are messengers of God, generally depicted with wings.

Left to right **The eagle** is associated with perception and insight, while the woodpecker helps find hidden patterns and coincidences. Birds' ability to fly connects them to the sky. In all cultures they have been thought of as providing a supernatural link between the Heavens and the Earth.

Left to right **Egyptian incense offering; Asian incense; Native American pipes.** Smoke is used in many cultures as a medium for carrying messages to the spirit realm.

Colour of the air element

Like water air is colourless and transparent. White symbolizes its purity and its ability to assume any other colour. Air is also often symbolically coloured yellow since it is linked to the Mind.

28 Living art

The aim of art is to represent not the outward appearance of things, but their inward significance.

ARISTOTLE, GREEK PHILOSOPHER
(384–322 BC)

> To say the word Romanticism is to say modern art – that is, intimacy, spirituality, colour, aspiration toward the infinite, expressed by every means available to the arts.
>
> CHARLES BAUDELAIRE,
> FRENCH POET
> (1821–1867)

HUMANS HAVE AN INCREDIBLE FACULTY; the ability to be creative. We are larger than ants but smaller than elephants; the ideal size for manipulating our environment and the elements. We can see shapes, colours and elements in action through observation of the environment, where they are visible literally and symbolically. Our imaginations let us pose questions and look behind images, to see the forces that structure and animate them. We can then create symbols using shapes, colours and metaphors to represent these abstract forces in our belief systems, art, ceremonies, religious sites and even in the organization of our cities. Complex symbolism is found within the geometry underlying our many artistic applications to try to achieve union with the underlying creative power of the Universe.

Sacred geometry is the use of geometry with intent as a tool or vehicle, invoking purpose. Around the globe scribbles on walls, rocks and artefacts have evolved into sophisticated systems representing an inseparable fusion of art, mythology, science, music, religion and "magic". As tools, geometry, colour and the Classical Elements can be used in a variety of ways in our creative expressions to commune with reality and expand the Mind.

Creating living art

Works of art are creations of the "art of living", none more so than simple, aesthetically pleasing pictorial symbols based on sacred geometric principles, where form has a function and holds, or sustains, energy. They have been created down the ages, in all belief systems, arts and architecture to give tangible form to the intangible order uniting everything. As microcosmic pictures of the macrocosm these images share elusive, timeless ideas and universal values of deeper truths. Via abstract form these pictures are like complex mathematical equations displaying the underlying nature of the Universe.

As an act of creation, a vast variety of images and structures can be produced. When we make and use these creations as tools with intent in Mind, then our creations receive life. Every part of these living works of art has a function and purpose if employed because of what they represent – shapes and patterns, colour,

Classical Elements, animal and plant symbolism, materials and media, location and layout. Human creations of this nature powerfully attract people as they are imbued with symbolism and intent, linking to our soul and subconscious.

Union with the absolute through sacred geometry

Yantra Yoga is the path of union with the Absolute through geometric visualization. It can also be known as Mandala Yoga. The shapes incorporated in a yantra are thought-forms that create patterns of energy. When used for meditation, geometric yantras are tools for focus, so that we can commune with these visual metaphors, to move beyond the boundaries of the physical planes. The purpose, method and intent necessary to use them to effect can take years of study.

Whatever the culture and geometric model used to represent the underlying nature of the Universe and our path toward union with the Absolute, they are all subtle, complex and rich with metaphors. Striking similarities cannot be discounted even though we may be removed by culture, language and time. But these models do contain visible clues that we can intuitively recognize; in shapes and numbers, colours and Elements. We need to transform our conscious abilities by cultivating a life based around such use of symbolism if we are to realize our own potential.

Above **When an apple is cut through its equator both halves reveal a pentagram shape at the core, with each point on the star containing a seed.**

Classical Elements in the magic pentacle
Nature magically mixes the forces of the Elements in her creations. Humans associate the pentagram (or pentacle) with magic and the Elements. It is widely used in diverse cultures and religions, originating as the symbol of the Goddess Kore and her sacred fruit is the apple (see above right). The Roma refer to the apple core as the Star of Knowledge.

Underlying creation of artwork
Artists employ 5 key tools in the structure of a piece of art: line, tone, colour, composition and perspective. 5 solids are used to create 3-dimensional forms; cube, sphere, cylinder, pyramid and cone. The Gateway to the Heavens, considered as art, has 5 parts to its structure; Circle, Square, Triangle, Cross and Spiral.

Below left to right **A line drawing with no perspective; a line drawing with perspective; a tonal drawing; a drawing containing line, tone, colour and perspective.**

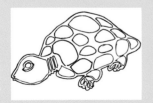

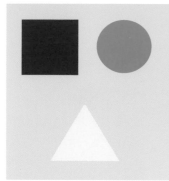

3 shapes 3 colours – Bauhaus School

At the German Bauhaus School of Art in the 1920s, Paul Klee and Wassily Kandinsky carried out a study of the propensity to ally shapes with certain colours, to determine whether there was a universal law of psychological relationship between them. A thousand postcards were sent out, asking people to colour a Square, Triangle and Circle with the 3 primary colours. A majority filled the Square with red (Earth and blood), the Triangle with yellow (intellect) and the Circle with blue (insight and spiritual truth).

Above **Buddhist monks making a sand mandala.**

Construction of a yantra

Making a yantra requires precise use of geometry, proportions and symmetry. There are even rites involving careful selection of the timing and materials, ceremonial purification, isolation and complete concentration through meditation and prayers. This is similar to the creation of Zen diagrams, where the intent of the Mind is of paramount importance, as are the way the brush strokes are applied and the simplicity of the form used.

CHINESE ELEMENTS, COLOUR, SHAPE AND DIRECTION

This is an example of Chinese interpretation of colour, geometry and the directions.

	Wood	Fire	Earth	Metal	Water
Colour	Green	Red	Yellow	Purple	Blue
Shape	Rectangle	Triangle	Square	Circle	Curve
Material	Wood, plants	Fire, light	Stone, clay	Gold, silver	Water, mirror
Direction	East	South	Centre	West	North

Right **Air is represented by a blue Circle; earth a yellow Square; fire a red Triangle; water a crescent shape, while spirit is the black oval shape.**

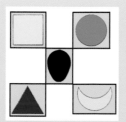

The tattvas

Tattva is a Sanskrit word meaning "thatness", "principle", "reality" or "truth". According to various Indian schools of philosophy, a tattva is a symbol that represents a feature of reality conceived as a characteristic of a deity. Although the number of tattvas varies, depending on the philosophical school, together they are thought to form the basis of all our experience. The Samkhya philosophy uses a system of 25 tattvas, while Shaivism recognizes 36.

Using sacred geometric works of art

Essentially the rituals for using any geometric models are the same. They involve the creation of an energy circuit, or link, between the model and the Absolute. Though intent is in the Mind, the energy of the image is cleared of negativity and fenced in so that it cannot return. Users then clear themselves of negativity and prepare their Mind further. The technology of geometry employed in geometric visualization is similar to the precision needed to set up an electrical circuit so that the current can flow through. The right mindset, or intent, when creating and using these images, is essential, otherwise the geometric patterns remain as pictures and a connection is not made. At the end the connection is reversed so that it becomes just a picture once more. Concentrating on the geometric shapes melds us with the model and beyond into ultimate reality. Just as the shapes facilitate creation of life so they act as aids to bring order to the Mind. In principle the image becomes so real within and without that it "lives", absorbing attention fully and integrating the viewer with all creation to show the essence of unity.

Above **Ornate geometric symbolism from the 6th-century *Book of Kells* (Celtic).**

Native American Medicine Wheel

Native Americans use a type of yoga, since they incorporate geometric patterns in their rituals of worship. The Medicine Wheel is an outward expression of an internal dialogue. The Circle represents life's never-ending cycle. The Four Directions focus on different aspects of life that have to be in balance for all to be well with the world.

North	White	Winter	Earth	Purity, wisdom, higher power, guidance
East	Yellow	Spring	Air	Flight, beginnings, dawn
South	Red	Summer	Fire	Passion, growing, vigour, youth
West	Black/blue	Autumn	Water	Emotions, endings, reflection, soul-searching

A personal Medicine Wheel can be made using fetishes such as crystals, arrowheads, seashells, feathers and animal fur or bones. The Wheel can be used to contemplate the balance of each aspect of life (self, family, relationships, life purpose, community, finances and health). The Wheel can help the individual gain new perspectives on life. During construction the person begins to sense what areas of their life are not in balance and where their attention is lacking. They continue working with the wheel after constructing it by sitting within its boundary in silence.

Above **A Hopi ceremony with a basic Medicine Wheel drawn on the ground (in white).**

29 Combining shape, colour and Element

Creativity is not the finding of a thing, but the making something out of it after it is found.

JAMES RUSSELL LOWELL, AMERICAN DIPLOMAT, ESSAYIST AND POET (1819–1891)

> Every artist dips his brush in his own soul, and paints his own nature into his pictures.
>
> HENRY WARD BEECHER, AMERICAN ABOLITIONIST AND CLERGYMAN (1813–1887)

IN THESE FINAL TWO CHAPTERS THE MESSAGES of the previous chapters are consolidated and presented as summary illustrations based on personal intuitive insights and interpretation of information on the meaning and purpose of geometry, light, sound and the Elements. In this chapter, the basic information linking shape, colour and musical notes is summarized in simple visuals.

The integration of shapes with specific colours and tones elicits subtly different effects, depending upon the context and intent behind their use and how they are employed as symbols. In this book they are regarded as participants in the process of creation and which combinations are appropriate for incorporation in the Gateway to the Heavens model in Chapter 30 (see page 224). In other contexts the use of symbols deliberately in creative expression, be it in art, commerce, beliefs or in science, will be imbued with the intent of their creator and users.

Warm colours, shapes and Elements

Taking the 7 colours that the human eyes can discern and placing red first as it naturally correlates with the Square (itself symbolic of the Earth), since the colour of blood is symbolic of the earth Element and life on the surface of Planet Earth. Our physical body comes from, and returns to, the Earth as well.

Red of the Earth and yellow representing the Circular Sun combine to make orange, which is used in the Triangle representing transformative, ordering fire and the physical aspect of Being. The heat of the Sun's yellow flames is tempered by the red of female Earth to make the orange flame that is used by man; since fire requires fuels of the Earth, such as wood and charcoal, and oxygen from the air to burn.

Air is also often symbolically coloured yellow since it is linked to the Mind. The physical air of the Earth is warmed by the Circular Sun. Red, orange and yellow are highly physical, warm colours and as such it is appropriate that they are allocated to the Square, Triangle and Circle, since they are representative of warm physical principles.

Combined symbolism of warm colours, shapes and Elements

From within our physical bodies on the surface of the Earth (red) supported by the solid foundations of the Square we look upward at the Sun, the yellow Circle and the source of physical life, shining light with all its colours so that physical Being may be revealed. In between red and yellow is the orange; a combination of the yellow light of the Sun and red of Earth. This can be symbolized by a Triangle graduated by these warm tones, with yellow at the apex. Beneath this Triangle of the Earth would be subterranean black.

Increasing density
Slower vibration

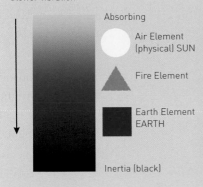

Absorbing

Air Element (physical) SUN

Fire Element

Earth Element EARTH

Inertia (black)

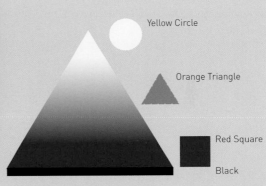

Yellow Circle

Orange Triangle

Red Square

Black

The Fire pyramid and Eye

The apex of the "Fire" Pyramid marks the Centre of the Square and Circle where the All-Seeing Eye resides and from where light radiates outward.

Above and centre **Apex of the Pyramid is the Centre of the Square.**

Above right **As a Circle, "Sun" light radiates out from the Centre of the Circle.**

Neutral green

Yellow light is stored by the green plants and used to produce colourless, physical air and vegetation, both of which sustain physical life. Blue, representing the cool sky and also water (linked to the dynamics of emotions), combines with warm yellow (Sun) to make balanced, healing green. In the middle of the colour spectrum green is neither warm nor cool. It acts a neutral fulcrum between the realms of heaven (spirit) and Earth (matter) and hence it is symbolically linked to the Cross. Unconditional love, compassion and healing correlate to the Cross and Heart (see page 33), which form a kind of moral, or judgemental, "gateway". At this neutral green gateway we face the trials of the Cross, questioning our thoughts, words and deeds.

Neutral green Cross

221

Far left and left **Lush greens surround us; calming and "cooling" what would otherwise be a harsh and hot environment. Green links us to the Earth through the food chain, trapping the energy of the light and converting it to food so that we can fuel our bodies.**

Right **Tonally, white and black are neutral grey when blended.**

Cool colours, shapes and Elements

Beyond green lie the less easily discerned cool, blue-based colours. We naturally associate these colours with water, sky, space, the heavens and the intangible concepts of the spiritual world, such as the spiritual body and ether (the spiritual breath prana, or Life Force). All these are essentially shapeless and colourless and their dynamics are more fluid, exhibiting Circling and Spiralling natures.

Blue is also an instinctive association to make with the Circle, as was discovered by the Bauhaus research (see page 218). It makes a natural association with cool, blue water and the blue sky just above the Earth. Feminine, watery emotions are expressed from our throat and in our voices as sound and in our watery sweat and tears. Indigo is also associated with the Circular vaults of the Heavens,

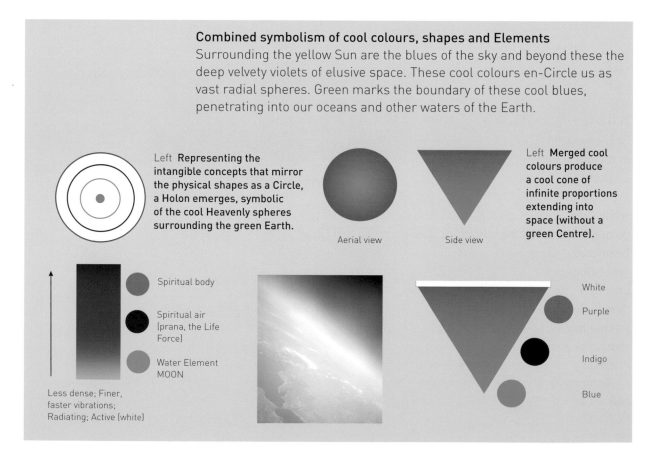

Combined symbolism of cool colours, shapes and Elements
Surrounding the yellow Sun are the blues of the sky and beyond these the deep velvety violets of elusive space. These cool colours en-Circle us as vast radial spheres. Green marks the boundary of these cool blues, penetrating into our oceans and other waters of the Earth.

Left **Representing the intangible concepts that mirror the physical shapes as a Circle, a Holon emerges, symbolic of the cool Heavenly spheres surrounding the green Earth.**

Aerial view   Side view

Left **Merged cool colours produce a cool cone of infinite proportions extending into space (without a green Centre).**

Spiritual body

Spiritual air (prana, the Life Force)

Water Element MOON

Less dense; Finer, faster vibrations; Radiating; Active (white)

White

Purple

Indigo

Blue

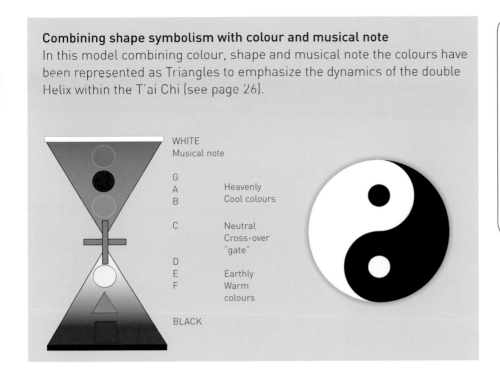

Combining shape symbolism with colour and musical note
In this model combining colour, shape and musical note the colours have been represented as Triangles to emphasize the dynamics of the double Helix within the T'ai Chi (see page 26).

WHITE
Musical note

G
A   Heavenly
B   Cool colours

C   Neutral
        Cross-over
        "gate"

D
E   Earthly
F   Warm
        colours

BLACK

> Art is born of the observation and investigation of nature.
>
> CICERO,
> ROMAN AUTHOR,
> ORATOR AND POLITICIAN
> (106–43 BC)

beyond the planet Earth, containing the stars and the domain of spiritual air. Red of the Earth, with her rich brown soil and cool blue waters, are the primary materials of our bodies, combining to make purple and blending the blue end of the colour spectrum with the red.

Balancing the material and spiritual

At the simplest level we have seen that red, white, and black are the 3 colours of the qualities of material nature. White is the upward, dispersing force of pure consciousness, containing all the rainbow colours, red represents the dynamic principle, while black implies inertia and the downward, absorbing force. Pure white light is veiled by darkness and increasing density as one goes down the planes of Being and the vibrations slow down, so that Beings of matter may form. Below red, below the Earth, is an absence of light, as in caves, and the condition before being born. The pitch black of death absorbing all colours represents unconsciousness. The T'ai Chi symbol (see page 26) is universally symbolic of this duality between black and white, each of which requires the other to show its true nature. At its mid-point is a balance of Yin and Yang. When the T'ai Chi symbol is placed beside the combined model of the cool and warm colours the dualistic personality of the heavenly and Earthly parts can be better appreciated.

Numerically the T'ai Chi represents 2 and the Duals. It also has within it 01 and 10 as the black and white dots within their opposite tone. Numbers 0, 1, 2 and 10 are the beginning and end, and between them hold the other 7 numbers (3, 4, 5, 6, 7, 8, 9) structuring and orchestrating reality

30 Gateway to Becoming

Now where there are no parts, there neither extension, nor shape, nor divisibility is possible. And these monads are the true atoms of nature and, in a word, the elements of things.

GOTTFRIED LEIBNIZ, GERMAN PHILOSOPHER AND MATHEMATICIAN (1646–1716)

Above **A very elemental Gateway to Becoming.**

AS A GROUP, THE CIRCLE, SQUARE, TRIANGLE, CROSS AND SPIRAL unite as one sacred geometric model, the Gateway to the Heavens, which is activated into a Gateway to Becoming when the combined symbolism of colour, sound and the dynamic metaphors of the Elements are added to it. Outwardly its appearance is simple, but inwardly it is complex since the subtle purposes of all the numbers (0, 1 to 10) and geometry are "hidden" within it. This unified, universal model reveals the nature of the highly creative holographic Universe through abstract symbolism. It serves as an object of contemplation that merges with our hearts and souls, cutting across all dimensions of reality and as a vehicle for personal creative expression that mirrors the creation of reality.

Adding Classical Elements and colour

Colours are used in conjunction with sacred geometry primarily for their symbolic significance and in practice to elicit effects. So colour symbolism is allocated to specific shapes accordingly, as we have seen in Chapter 29 (see page 220). Since the Gateway to the Heavens model establishes the frameworks of Space-Time-Being and facilitates the process of becoming ("to become" a rock, tree, spiritual being, planet or gas molecule) by adding the dynamics of the Classical Elements and colour metaphysically and symbolically it then represents the creative process and the condition of Being.

There are many ways of adding colour and the Elements to the basic Gateway to the Heavens model as part of individual creative expression, using the imagination. The illustration on the cover of this book is one example. Most cultures assign a physical Element to each Direction, but in the Gateway to Becoming they are incorporated within the shapes and their colours. In the illustration on the cover of this book visual images have been assigned to each corner of the model in corresponding pairs of male and female, as in traditional alchemy. To denote the spectrum of colour symbolism and musical notes a Circular rainbow is shown. Formed by water and light, the rainbow also embodies the synergy of the

FINAL STAGE TO BUILDING UP THE GATEWAY TO BECOMING

The components of the Gateway to the Heavens model with the Element and colour added

Question	Shape	Centre	Active word	Aspect	Element	Colours
When	Circle	Moment	Now	Time	Air, water	Yellow, blue
Where	Square	Point	Here	Space	Earth	Red
Who	Triangle	Eye	I am	Being	Fire	Orange
Why	Cross	Choice	To	Centre	(Neutral)	Green
How	Spiral	Sharing	Because	Life force	Ether	Indigo

Through their inclusion all the 7 colours, and with them the 7 musical notes, become an integral part of the model.

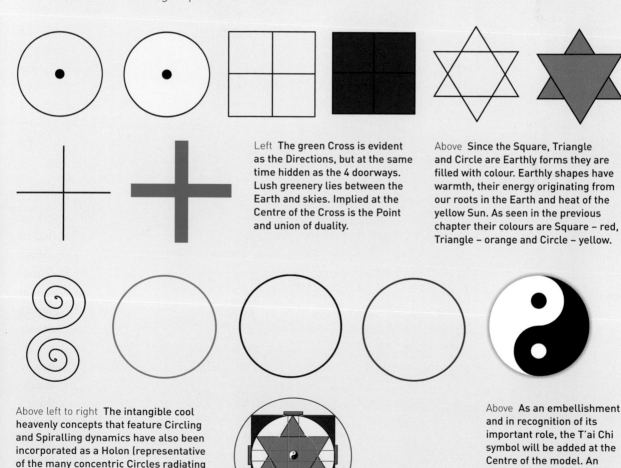

Left The green Cross is evident as the Directions, but at the same time hidden as the 4 doorways. Lush greenery lies between the Earth and skies. Implied at the Centre of the Cross is the Point and union of duality.

Above Since the Square, Triangle and Circle are Earthly forms they are filled with colour. Earthly shapes have warmth, their energy originating from our roots in the Earth and heat of the yellow Sun. As seen in the previous chapter their colours are Square – red, Triangle – orange and Circle – yellow.

Above left to right The intangible cool heavenly concepts that feature Circling and Spiralling dynamics have also been incorporated as a Holon (representative of the many concentric Circles radiating from the Centre).

Right The completed Gateway to Becoming model.

Above As an embellishment and in recognition of its important role, the T'ai Chi symbol will be added at the Centre of the model. An Eye, as used before (see pages 115–116), is another possibility, but a simple Point would suffice.

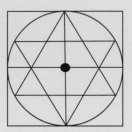

Gateway to the Heavens

A simple Holon within the Gateway to Becoming

A simple Holon exists as part of the Gateway to Becoming (see diagram below right). Like the Spiral and the circuits of the yantra, this intangible Holon is not necessarily physically evident in a drawing. But, when included, notice how water (blue Circle) circulates within Being (orange hexagram). Also the 2 Circles representing spiritual air (indigo) and physical air (yellow) coincide. Spirit (purple) and the original vessel encircles the geometric principles within it, containing and protecting them within its boundaries.

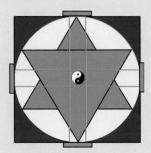

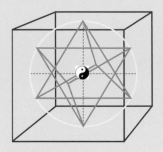

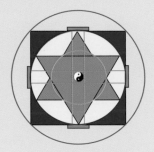

Gateway to Becoming

Orange Triangle of Being and blue Circle of water. Additionally, rainbows are generally symbolic of the bridge between heaven and Earth and of the fertile imaginations of the Mind. According to Japanese mythology the creators of the world descended on a "floating bridge of heaven" to create land out of the sea of chaos. Australian Aborigines believe the rainbow serpent is the creator of the world and all Beings as it brings fertility to Earth by creating the rain from heaven.

Sound activation of geometric models

Dr Hans Jenny (see page 140) noticed in his research in cymatics (sound wave phenomenon) that the vowels of the ancient languages of Hebrew and Sanskrit assumed shapes that were the same as the written symbols for these vowels, something that does not occur with the sounding of letters in English. Music similarly exhibits hidden geometric codes. Hindu yantra each have an appropriate mantra, which is a Sanskrit sound syllable unique to each yantra. Sanskrit letters embody the concept of Dualism being divided into male and female. Every consonant having male energy is articulated with a vowel embodying female energy, so the female gives life to the static male and expression to words and sound.

These mantras are not to be viewed as words for functional speech or as musical notes, but as subtle vibrations of sound complementing each yantra. Geometric image and sound are interchangeable and equally important, which is vividly expressed as "seeing sound and hearing form", as found in cymatics research. Essentially mantras are thought-forms, expressions of the Mind that exert an influence over the geometric yantra to release and intensify its power.

Colours of the Yantra

Even ancient symbols such as the Sri Yantra have been re-interpreted using different colours to elicit varying effects. Traditional earthy, muted tones are used in the Tibetan Yantra (below right). In it you may notice the diagonal division of the Square and the allocation of colours to the Directions: white–north, red–south, yellow–west, green–east.

Rainbows

Rainbows are formed by light refracting as it enters a drop of spherical water and then reflecting inside the droplet at an angle of 42° (as shown in the diagram below). Second rainbows form if the light is reflected twice inside the droplet.

Clouds of water droplets spread out in three dimensions and reflect the light toward you in the shape of a cone with your eye at the tip (see René Descartes' rainbow below centre). This is why the image of a rainbow forms as a complete Circle, but with its bottom section cut off by the horizon of the Earth. Every rainbow is the same size, although we only see varying portions of the Circle depending upon the height of the Sun.

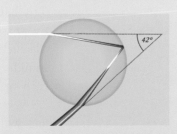

Above **Light refracting and then reflecting inside a water droplet.**

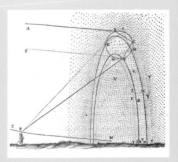

227

Top and above **While rainbows are seen with your back to the Sun through waterdroplets, Sun dogs appear as a thin rainbow halo around the perimeter of the Sun and are caused by ice crystals in the atmosphere. The term "dogs" refers to 2 light balls on each side of the Sun within the halo.**

> The lamps are many but the light is One.
>
> RUMI,
> PERSIAN POET, MYSTIC,
> THEOLOGIAN AND
> JURIST
> (1207–1273)

Intent and vibration are essential to their effectiveness. They can only be repeated with the right mindset; in other words the correct intent in the thought. They must to be correctly pronounced, merged, accented and intoned.

Geometric shapes of AUM

AUM/Om is the most widely known and powerful seed mantra, whose form is also a yantra, the Sri Yantra. Although there are believed to be many hundreds of ways of pronouncing AUM/Om, each having a different effect on matter and spirit, when it is correctly toned into Hans Jenny's tonoscope (page 147) it forms the Circle. As the chanting continues further concentric Squares and Triangles and finally a Sri Yantra is produced as the last sound dies away. AUM reveals all and contains all. The primal sound vibrations of AUM/Om contain the geometric blueprint of reality, as used across all the Planes of Being underlying every form and every principle of Nature. Om/AUM is also associated with white light, containing all the colours and the source and basis of life.

Examples of seed mantras

Like yantras, mantras are categorized according to their purpose. For example, those inducing trances are used with yantras for enlightenment. Most effective are seed mantras (*bijaksara*), which are single syllables comprising combinations of letters that contain the sum total of divinity.
- KRIM is the seed character of Kali, comprising K = Kali, R = Brahma, I = Maya, M = Mother of the Universe and the bindu.
- SRIM is the goddess of good fortune, derived from S = abundance, R = wealth, I = satisfaction, M = unlimited.

Sri Yantra and A HA

Since the Sri Yantra is such an important yantra it usually has all the Sanskrit letters, vowels and consonants incorporated in the angles of the pattern. A, the first Sanskrit letter, represents Siva. When pronounced correctly it sounds like "Ha" and the sound of the last letter that represents Sakti. Through sound their interdependence is conveyed: one moving, the other still. The combination of A and Ha is Circular, symbolic of beginning and end, embracing all the letters held within them. Together they represent all creation and its vibration as sound. The Sanskrit letters sometimes inscribed in a Spiral on the Sri Yantra are the "gross" aspects of sound that occur with the union of Siva and Sakti and the energies of the 5 Elements (see page 146).

The extended seed mantra AHAM, means "I" and comprises A = Siva, HA = Sakti, M = union. Therefore their union creates identity.

Sri Vidya

The Sri Yantra is also used in the instruction of the Sri Vidya, which requires the use of the 15-syllable Sri Vidya Mantra. The ritual practice of this mantra is conveyed orally according to the level of spiritual awareness. Comprising 15 (3 x 5) seed vibrations, believed to hold the force of enlightenment, the Sri Vidya mantra is also used in conjunction with the Yantra of Nine Triangles (3 x 3) known as the Nava-yoni Chakra. The seed vibrations are divided into 3 groups, inscribed in the Triangles around the central bindu. These vibrations represent all the hierarchies and Planes of the Universe. One of these seed vibrations, Cit, is purported to represent the nature of pure consciousness. The use of this seed vibration, when intoned correctly, can bring about direct comprehension of the Sri Yantra – Sri Vidya Mantra thought vibration.

Above **His Holiness Dagchen Rinpoche's hand is held by monks as he draws lines that close the Hevajra Mandala, after the empowerment (Tharlam Monastery of Tibetan Buddhism, Boudha, Kathmandu, Nepal).**

Embodiment of 5 in AUM

When written, AUM is broken down into 5 parts that directly correlate with the 5 Elements and 5 senses (see page 146). Siva's eternal energy is made evident in 5 activities and his mantra also has 5 letters. Each of the 5 seed mantras, the single syllables that are also symbols of an attribute or quality of the Divine, may also be related to each of the 5 parts of AUM.

Doors into the Gateway

A gateway marks an entrance. It never vanishes, but remains to protect and shelter us. When we go through, a symbolic death and rebirth as we "change" perspective, awareness or concepts of reality, occurs. The shape, or combination of shapes, denoting the gateway is symbolically relevant. The Gateway to the Heavens, like the Sri Yantra, has an enclosing Square including 4 entrances in each side facing the 4 cardinal Points. Each is a door into the Square domain through which the Mind may enter or leave during contemplation. These 4 doors are at the ends of the arms of the Cross, delineating the Central Point of the model common to spirit and matter and hence the Gateway between Earth and Heavens. According to yantra philosophy, before we can reach the Centre, we must resolve the ways of each of the 4 Directions. Once we get to the Cross, learn its lessons and take on board its wisdom we can enter the Central Gateway.

The Centre, the pivot

To facilitate creation the unity of the Point is split into 2. Duals are fundamental if creation is to take place through the interplay of opposites. Our Mind is capable of resolving this duality the closer it moves to the Centre. So, to reach the Centre we face the "battle" between light and dark forces that are in eternal

> Yes, Love indeed is light from heaven;/
> A spark of that immortal fire/ With angels shared, by Allah given/
> To lift from earth our low desire.
>
> Lord Byron,
> English poet
> (1788–1824)

> Rise above
> sectional
> interests
> and private
> ambitions...
> Pass from
> matter to
> spirit. Matter
> is diversity;
> spirit is light,
> life and unity.
>
> MUHAMMAD IQBAL,
> INDIAN MUSLIM POET
> AND PHILOSOPHER
> (1877–1938)

contradiction. These are in the Choices of the Cross we make in every Moment; in thoughts, words and deeds. The Centre acts as a pivot around which the other shapes form, like a holographic seed for reality. Physicist Dr David Bohm maintains that the primary physical laws cannot be discovered by a science that attempts to break the world into its parts. He writes about an "implicit enfolded order", which exists in an unmanifest state and this order is the foundation of physical reality. When reality manifests physically he calls this "explicate unfolded order", in which parts are interconnected irreducibly. Their relationships depend on the Whole. When we move into sacred images we can experience the fusion of all Planes and reach the Point where individual, symbols and Universe are fused. The challenge is to merge with the Centre, the ultimate form that cannot be reduced. From the Central Point the Life Force unwinds and rewinds the energy of consciousness into Forms of Being, concentric Circles expand and contract around the Moment.

Knowledge of the light

Since the bindu is symbolically placed in the Centre of the forehead on the the Third Eye we know the Centre can be reached by opening and focusing the 3 Eyes. By gathering pieces of "light", or knowledge, we can emerge from the "darkness" of ignorance and escape physical bondage. All 3 Eyes sharpen our focus of attention and raise awareness of the Choices we make at the core of our existence to higher and higher levels. This process of increased focus is represented in the Sri Yantra as 9 concentric Circles, centred on the bindu and as the Holon in the Gateway to Becoming (see page 226). The aim is to move the Mind from outer Circles to each inner Circle, until the Centre is reached. These 9 closed Circles correspond to each of the planes of Being and mark stages of apparent ascent, which is not just a culmination, but also a fusion of all the Planes of Being where the individual, symbols and the Universe are one. When making Choices and experiencing contradictions in life, coloured emotions are felt in our Hearts. The Heart is linked to the Cross and when the Centre is reached and experienced momentarily, pure light flows through us as unconditional Love.

Becoming and fragmentation, yet all is One

The Gateway to the Heavens model establishes a framework that is transformed into a Gateway to Becoming through the incorporation of colour symbolism for light and the fabric of our Being. Sound acts as the medium of creation, while the ingredients of life are shown by the dynamic metaphors of the Classical Elements embodying number 5 for the Life Force and the active process of Creation. All Beings are fashioned out of the same ingredients and the same geometric rule set, yet generating variety through pattern-sharing. Everything and anything is possible. Everything on all the dimensions of reality is facilitated by geometry. It is incredible that these underlying mechanisms have been put into place so that reality can come into existence. This is true genius that we are exploring with our Minds and expressing though our symbolic systems.

Our sensory experience of reality is also amazing, as is the fact that we can share our experiences with each other. We are like participants of a flickering film with our unique roles in a production of immense proportions and complexity. Each life is a precious gift to experience the Duals in every waking Moment and the emotions that come with them. Our ability to create our reality through our Mind and to influence the course of events through our intent is a powerful facility. It is also an ability to be used while mindfully aware of our conscience and of the consequences that will impact every other Being. Everything we see, all Beings in time and space, are part of the Whole. Our Eyes craft a window through which we look onto the shimmering image of this reality of which we are part. Our Eyes and all the other Eyes of sentience are a part of the Absolute made conscious. Any feeling of separation, isolation and individuality generated by looking out of our window is an illusion; an illusion generated and revealed by light.

Everything created also has a Centre that is connected by the umbilical cord of the Life Force to the common Centre of the Absolute, whole, complete Being, of which we are part. The Many are One and the experience of this intimate connection with all other Beings can be made by using geometric models like the Gateway to the Heavens. On a cosmic level, when the Gateway to Becoming is implemented in reverse the Many go back to Being One in the Centre. Matter dissipates and then involves, drawn into the darkness once more to be born again as light and a new reality.

I AM, HERE, NOW

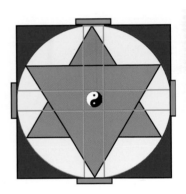

At the Centre of the Gateway to the Heavens in my Mind "I am, Here, Now".

I am sentient and part of reality.

"I am" a dualistic Being of spirit and matter, created by the thoughts of my Mind, present in the Moment, Here.

And in every Moment "I am" reminded by the Duals of the Cross that "I Am, Here, To..." make choices. "Because" every choice is a thought and my thoughts have intent.

My thoughts precede my words and deeds.

My thoughts create my reality and my intent is part of my reality and your reality.

"I am" one and also part of All, or the One, so everything I think, say and do affects every other form of Being.

"I am" a Being of light, formed through sound, characterized by the Elements.

"I am" sustained, nourished and animated by the Spiral Life Force in a perpetual, eternal life cycle of all Being.

Life is a colourful sensory experience of learning, felt through the emotions in my Heart Centre, giving my life meaning and purpose.

My Being's song vibrates, rejoicing, joined with the other voices and music as part of the harmony of the Universe.

"I am, Here, Now"

K. L. French

Glossary

Alchemy The term given to a form of medieval chemistry that combined science with mysticism. Alchemy was also practised by the Ancient Egyptians and Taoists in China.

Apex Highest point or summit. The apex of a Triangle is opposite its base.

Archimedean Spiral A Spiral within the distance between each coil remains the same. (*See* Helix.)

Archetype, archetypal First perfect type or form, the original model, from which copies are made.

Atman "Self" in Sanskrit, referring to your true self beyond physical parameters.

AUM (Om) (Hinduism) The most ancient sound symbol for the original source. It acts as a non-specific name believed to be the original sound of creation and is frequently used in mantras.

Bindu (Hinduism) Sanskrit term meaning "point" or "dot"; the smallest created thing and the central Point of the Universe.

Brahma (Hinduism) Four-headed Hindu god responsible for creation and one of the Hindu Trinity (the other gods being Siva and Vishnu).

Brahman (Hinduism) Impersonal absolute or supreme deity. The universal soul or energy that is present in all things manifest. Also a member of the highest Hindu caste, originally priests.

Caodaism Monotheistic religion, started in Vietnam (1926), that is a blend of various other beliefs.

Cardinal points Four main directions of the compass – north, south, east and west.

Cerebral cortex Located at the front of the brain behind the forehead, the area of the brain most associated with the generation of conscious awareness.

Cerebrum Main part of the brain in the front of the skull, which integrates complex sensory and neural functions.

Chakra (Shakra) Sanskrit word meaning "wheel" and "centre". A concept originating from Hindu texts, referring to "force centres", or rotating wheel-like energy vortices, which are believed to exist in the surface of the etheric body of all living beings. Each chakra has a central point through which it draws energy from the Sun/prana/chi and circulates it through the body and aura. There are seven major chakras situated up the central line of the human body; Root/Base, Hara/Sacral, Solar Plexus, Heart, Thyroid/Throat, Pineal/Brow and Crown.

Chi (Taoism) Literally "air", "breath" or "gas", but most often translated as the "life force" and comparable to the Hindu concept of prana. One of the life-giving Three Pure Ones shared by humans and the cosmos (the other two are *jing* and *shen*).

Codex Book in the format used for modern books, with multiple gatherings of paper or vellum bound together in a cover.

Commune To talk together, a shared experience and close connection.

Copernican Term describing theories attributed to Nicolaus Copernicus, the French 15th-century Renaissance astrologer.

Cosmic Relating to the cosmos, space.

Cosmic womb The downward Triangle that surrounds circles for Siva and Sakti is known as the Primal Triangle, representing the "cosmic womb" of the creative female Sakti.

Cyclic Of or in recurring cycles.

Cymatics Study of visible sound and vibration.

Decad (Greek) Number 10 and its principles. The word comes from *deca*, meaning "having ten" and *dachus* for "receptacle", since 10 contains the whole.

Diatonic Describes a musical scale comprising intervals of five whole steps and two half steps.

Digit Any of the numbers from 0 to 9 in the Arabic system; also a finger or toe.

Dimension Measurable aspect of a feature of reality (eg weight, length, duration).

Divine Term describing God and the sacred.

Dodecahedron Any polyhedron with twelve flat faces. These faces are regular, as is seen in a Platonic solid dodecahedron.

Dual Having two parts.

Dualistic Being of two parts.

Dyad Greek philosophers' principle of "twoness".

Ego Part of us that senses and reacts to reality and recognizes ourselves as an individual; the conscious part of the personality.

Electromagnetism One of the four fundamental interactions, the force that causes the interaction between electrically charged particles; the areas in which this happens are called electromagnetic fields. Also the force that holds electrons and protons together inside atoms.

Elements (metaphysical) Earth, wind, fire and water, believed to be the basic, indestructible ingredients of the Universe. A fifth Element is believed to be the force that animates life.

Element (chemical) Pure chemical substance consisting of one type of atom differentiated by its atomic number, which is the number of protons in its nucleus. For example, carbon, nitrogen and oxygen.

Elixir of life Liquid giving everlasting life.

Enlightenment Sudden realization of the illusion of physical reality and the existence of spirit; comprehension of the nature of one's own life and the nature of the Universe.

Ennead Number nine, group of nine and "the horizon".

Enneagram Geometric figure with nine points and sides.

Ether From the Latin meaning "to burn or glow", it is the fifth Element, which transcends and unites all the other four metaphysical Elements.

Event Anything that happens. In mathematics it is the outcome of a trial in probability.

Evolution Slow, continuous process of change in characteristics of organisms from one generation to the next.

Fetish(ism) Religious or mystical qualities attributed to inanimate objects.

Fibonacci Series Unique numerical series that starts with 1, adds 1 to itself, and each number thereafter in the series is the sum of the two prior numbers.

Fractal Rough or fragmented geometric shape that can be split into parts, each of which is (at least approximately) a reduced-size copy of the whole.

Geocentric (Universe) Geocentric theory (in Greek *ge* means "earth"), also known as the Ptolemaic system, maintained that the Earth was the centre of the Universe and that all other objects orbit around it.

Geomancer Expert in divination, who interprets markings on the ground or the patterns formed by tossed handfuls of soil, rocks or sand.

Geomatria Ancient version of numerology involving giving numerical values to words or phrases.

Geometry Area of mathematics studying the properties (such as lines, surfaces and solids) of figures and shapes in space.

Glyph A symbol that functions as a pictogram/ideogram or part of a writing system, such as a syllable or diacritic.

Harmonic Overtone accompanying a fundamental tone at a fixed interval, produced by vibration of a string or column of air in a fraction of its length.

Heliocentric or heliocentrism Astronomical model in which the Earth and planets revolve around a stationary Sun at the centre of the solar system.

Helix Three-dimensional version of the Archimedean Spiral often referred to as a coil, as found in a DNA molecule and screws/bolts.

Hertz (Hz) The International Standard for the unit of frequency of a wave.

Hexagram Geometric figure formed by two intersecting equilateral Triangles. Their angles coincide with the angles in the hexagon.

Hexagon A polygon with six straight edges and six vertices.

Hexahedron Another name for the geometric solid more commonly known as the cube.

Hierarchy Organization with grades and classes arranged one above the other in ascending rank, authority or importance.

Hieroglyph Picture or symbol used to represent a word or sound (for example, Ancient Egyptian).

Holography Photographic technique that allows the light scattered from an object to be recorded and later reconstructed so that when a camera or an eye is placed in the reconstructed beam an apparently three-dimensional image of the object is seen, even when the object is no longer present.

Holographic principle Property of quantum gravity and String Theory that states, "The description of a volume of space can be thought of as encoded on a boundary to the region, preferably a light-like boundary like a gravitational horizon". The principle suggests that the entire Universe can be seen as a two-dimensional information structure "painted" on the cosmological horizon.

Holon Expanding concentric Circles from a central Point and then contracting back to that original central Point.

Horus (Egyptian) Ancient hawk-headed god of the sky, war and protection.

Hypercube Square with planes that are hyperbolic or ellipses.

Hypersurface Generalization of the concept in geometry of the hyperplane, which is a generalization of the plane into a different number of dimensions.

I Ching (Yi Jing) Also known as *The Book of Changes*, an Ancient Chinese book of divination and one of the five classics of Confucianism. Written around 3000 BC by Fu Xi and later updated by Wen Wang in 1122 BC.

Ideograms Language based on patterns with ideas attached to them (for example, Chinese and Japanese).

Immutable Not changing; unchangeable.

Infinite Boundless, endless in space or time, greater than any quantity.

Intent Purpose behind one's Mind at the moment one thinks, speaks or carries out an action.

Kabbalah (Judaism) Doctrine associated with the esoteric aspect of Rabbinic Judaism.

Kali (Hinduism) Goddess and mistress of life, death and rebirth. She governs the Wheel of Becoming, or the Wheel of Reincarnation.

Karaïte Distinct Jewish movement characterized by the recognition of the Tanakh alone as its supreme legal authority.

Karma Sanskrit word meaning "force of action" coming from the root *kri*, "to do". The Law of Karma refers to how an individual's actions affect their future rebirth since every deed or moral act has consequential results.

Koran (Islamic) Central religious text of Islam revealed to the prophet Mohammed through the Angel Gabriel from God.

Labyrinth Symbolic of the journey to discover your Centre. An aid for learning about the spiritual path with a single, unambiguous route to the Centre and back.

Latent Present or potential, but not evident, manifest or active.

Latin Cross *Crux immissa* or *crux capitata*. This Cross has a longer vertical line, with shorter horizontals more than half-way up.

Lemniscate Double-looped infinity symbol commonly used today.

Lingam Phallus or phallic object, symbolic of Siva.

Macrocosm Large-scale picture of the Universe. (*See* Microcosm.)

Mandala (Buddhism) Pictorial, geometric representation of the Universe used for yogic meditation. Mandala comes from the Sanskrit, meaning Circle.

Mandorla (Hinduism) Two Circles over-lapping, representing the symbolic "vulva".

Manifest Clear or visible to the eye or mind.

Mantra (Hinduism) Word or verbal formula that is recited. A mantra is often used to aid meditation.

Matrix (mathematics) Regular array of numbers, symbols or expressions in rows and columns that is treated as a single quantity. The individual items in a matrix are its elements or entries.

Matter That which has mass and occupies space (opposite to spirit).

Maya (Hinduism, Buddhism) Principal deity that manifests, perpetuates and governs the illusion and dream of duality in the phenomenal Universe. Maya creates the physical world, which is seen as a "waking dream".The goal of enlightenment is to see intuitively that the distinction between the atman and the Universe is false and is the result of an unenlightened perspective.

Medicine Wheel (Native American) Constructed by laying stones in a particular pattern on the ground. Most medicine wheels follow the basic pattern of having a Centre of stone(s) with lines of rocks radiating out like "spokes" to an outer ring of stones.

Metaphysical Of metaphysics, based on abstract reasoning; supernatural; visionary.

Microcosm World or Universe in miniature form or representation.

Monad A term that, according to the Pythagoreans of Ancient Greece, means "divinity", or the first Being, or the totality of Beings that is One and without division.

Morphology Study of the structure of form.

Myriad Ten thousand of; great number of.

Mysticism Way of directly experiencing the Absolute through heightened awareness.

Nanomenter (nm) Unit of spatial measurement. 10 to the -9th metre, or one billionth of a metre.

Nirvana (Buddhism) Goal of Buddhism; the extinction of individuality and absorption into the Universe, never to be re-born. This liberation is achieved through the eradication of greed, hatred and delusion, which tie us to the indefinitely repeated cycles caused by karma. (*See* Karma.)

Occult Secret, esoteric knowledge beyond the range of ordinary knowledge.

Octagon Polygon with eight sides and angles.

Octahedron Solid figure contained by eight plane faces and usually eight Triangles.

Oscillating Repetitive variation, typically in time, of some measure about a central value (often a point of equilibrium), or between two or more different states.

Ouroboros Circular snake with its tail in its mouth, representing eternity, timelessness.

Osiris Egyptian god of the afterlife, underworld and the dead whose name can mean either "place of the eye" or "the one enthroned".

Pagoda Sacred building, especially a tower, which is usually of pyramidal form and commonly found in India and China.

Pentacle Sometimes used interchangeably with pentagram; however pentacles are amulets employed in magical invocations on which the symbol of an energy or spirit has been drawn. The pentagram star with five points (or Seal of Solomon) is the most common symbol used, but others may be used, such as the Star of David and Seal of God (a star with seven points).

Pentad Pythagorean term for the quantity of five.

Pentagram Geometric shape of a five-pointed star drawn with five straight lines.

Pentagon Any five-sided polygon.

Pentagrammaton Five-letter divine name Yahshuah and the Hebrew name of Jesus.

Perfect number Number equal to the sum of its divisors. For example, 6 = 1 + 2 + 3.

Petroglyph Image created by removing part of a rock surface by incising, picking, carving and abrading usually associated with prehistoric works – for example, human-like figures, animals, birds and abstract geometric forms such as Circles, Spirals, dots and lines.

Phi 21st Greek letter. The symbol Φ is also used to represent the Phi Ratio (or Golden ratio) $\Phi = 1.6180339....$

Phoenix Mythical bird that spontaneously combusts or burns itself on a funeral pyre, to be reborn again from its own ashes to live through another cycle.

Photon Finite "packet" of energy.

Phytoplankton Microscopic plant-like organism at the base of the marine food chain.

Pineal Pine-shaped body or gland located in the head at the back of the third ventricle. It secretes melanin.

Pituitary Major endocrine gland located in the head at the front of the third ventricle. It has a major influence on growth and bodily functions.

Plane Flat surface of two dimensions. A straight line joining any two points lies in a plane.

Platonic solids The five regular solids – tetrahedron, cube, octahedron, dodecahedron, icosahedron.

Polygon Geometric figure with many sides, usually more than four.

Polyhedron, polyhedral Geometric solid in three dimensions with flat faces and straight edges.

Prana (Hindusim) The Life Force, or spiritual air, and Dual to physical air.

Quantum (physics) Discrete unit of energy proportional to frequency of radiation.

Refraction Occurs when light changes its speed at the boundaries of different materials.

Sakti (Hinduism) Mother goddess comprising the entire pantheon of female forms; the virgin, fertility, warrior and nurturing mother.

Samsara Literally the "continuous flow

(of life)", the cycle of reincarnation within Hinduism, Buddhism, Bön, Jainism, Sikhism and other Indian religions.

Sanskrit Ancient language used for many of the ancient Indian texts.

Simplex Generalization of a Triangle or tetrahedron to n-dimensions in geometry.

Siva (Shiva) (Hinduism) God of destruction, cosmic consciousness and the non-manifest Point of stability underlying all phenomena.

String Theory Mathematical theory developed in the 1980s to explain the properties of elementary particles and the forces between them in a way that combines relativity and quantum theory. It describes quantum particles as vibrating strings. If the string vibrates at a different frequency it turns the particle into something else. This theory predicts that there should be ten "dimensions", six more that the current known four.

Symbol 1. Something that stands for or denotes something else. 2. Printed or written sign used to represent an operation, quantity, quality or relation (as in mathematics or music). 3. (Psychology) Object or image unconsciously used to represent repressed thoughts and feelings.

Symmetry 1. Sense of aesthetically pleasing proportions and balance that reflects beauty and perfection. 2. Precise concept of balance or "patterned self-similarity" that can be proved according to the rules of a system such as geometry.

Tao (Taoism) Path, the Way that is the universal principle responsible for the creation of all things, the "non-being", which is the basis for all "being".

Tessellation Pattern of plane figures that fills the plane with no overlaps or gaps.

Tetragrammaton Four Hebrew letters that refer to YHWH (Yahweh), the name of the god of Israel mentioned in the Hebrew Bible.

Tetrahedron Meaning "four faces", a solid with four identical Triangular faces. A pyramid with a Triangular base and the only three-dimensional shape whose vertices are the same distance from each other.

Tetraktys Greek for "fourfold". A Triangular model made of ten dots arranged in four rows of 1, 2, 3 and 4 and a very important symbol in Pythagorean philosophy and mysticism.

Toroid (mathematics) Doughnut-shaped mathematical figure formed by revolving a plane geometrical figure about an axis external to that figure, which is parallel to the plane of the figure and does not intersect the figure.

Transmutable Having the ability to transform, to change state.

Transmutation Change of one form, nature, substance or state into another.

Trefoil Graphic form made out of three overlapping rings that is used in architecture and Christian symbolism. Also used for other symbols of three-fold shape or arranged in three lobes, such as a knot.

Uniform material Material, such as water, having a uniform structure and consistency.

Unity State or quality of "being one"; single whole or totality, combining all its parts into one.

Universe "One turn" in Latin. All existing things; the whole of creation and the Creator.

Upanishads (Hinduism) Mystical texts of speculative philosophy, dating from about 600 BC. *Upanishad* comes from the Sanskrit *sad* and "to sit", and the two prepositions *upa* and *ni*, meaning "under" and "at".

Vedas (Hinduism) Ultimate canonical authority of sacred knowledge for Hindus. The oldest of the four Vedas is the Rig-Veda dating from between 1300 and 1000 BC.

Vesica piscis Latin meaning "bladder of a fish", an oval shape formed from the overlap of two identical Circles through their centres.

Vortex Any spiralling motion with closed streamlines.

Vitruvian Ideas attributed to Vitruvius, the Roman writer and architect who wrote the treatise *De Architectura* (1st century BC).

Waning (Moon) Phase of the Moon between full and new state; having progressively less of the surface illuminated.

Waxing (Moon) Phase of the Moon between new and full state; having a progressively larger part of the surface illuminated.

Yantra Sanskrit word for "instrument" or "machine", associated with elaborate and precise geometric patterns that are considered to be "thought forms" creating patterns of energy.

Yoga (Hinduism) Physical, mental and spiritual discipline of Hindu philosophy teaching a practical means of spiritual tranquillity and enlightenment.

Zen School of Mahayana Buddhism The word can be traced to the Sanskrit word *dhyana* meaning "meditation" or "meditative state".

Index

A

acoustics 132, 139, 182
air 213–215 *see also* Classical Elements
alchemy/alchemists 88, 90, 91, 92, 93, 94, 97, 100, 101, 120, 178, 203, 207, 210, 211, 214, 224
alphabet 87, 142, 143
    sacred 144
"amen" 144
amplitude 140
angels 39, 75, 129, 214
anti-matter 166
archetypes 14, 123, 142, 143
Archimedean screw 51
architecture 15, 16, 48, 52, 182, 185, 198
armillary sphere 15
art 15, 16, 17, 52, 71, 182, 185, 198
    living 217–219
Asian Indians 14, 142
Aspects of Reality (space, time, Being) 112–116, 199
    and the Gateway to Becoming 225
astrology 20, 93, 100, 203
atmospheric rippling 136
atomic numbers 102
atoms 27, 62, 96, 103, 106, 111, 120, 143, 160, 161, 163, 165, 167, 168, 187, 190, 210
AUM 144, 145, 146, 147, 228, 229
Aurora 138, 162
Australian Aborigines 14, 142, 226
Aztecs 39, 54

B

Babylonians 88, 167
balance 24, 26, 53, 64–65, 95, 96, 150, 208
baraka 49
beat 149
beehive 53
Being, Aspect of 112, 114
Big Bang 129, 132
Birkeland current 138
boundary 17, 23, 38, 109, 110, 111, 119, 190, 192, 197, 201, 219, 226
brain 23, 80, 81, 82, 86, 87, 134, 149, 197

breath/breathing 129, 142, 144, 149, 214, 222
Buddha 87
Buddhism 60, 84, 88, 177, 178, 210
Buddhist
    monks 210
    temples 75, 204
Buddhists 79, 144, 176, 196, 198

C

Calabi-Yau shapes 121
calendar 54
calligraphy 144
candle flames 196, 210
Caodism 84
carbon 103, 106, 200
carvings and scratchings 11
caves 202
Celts/Celtic world 14, 36, 178
chakras 186–187, 229
chants/chanting 135, 142, 143, 152, 153, 156, 212
chi/qi/ki 49, 101, 214
Chinese 88, 142, 177, 178, 211
Chladni figures 140
Christians/Christianity 39, 41, 57, 75, 84, 176, 177, 178, 179, 186, 211
Circle and the Centre 20
Circle, unity and the Duals 18–27
circles of humanity 22
circles and spheres 19
Classical Elements (fire, earth, air, water, ether) 7, 102, 122, 187, 200, 216, 224, 230
    as metaphors 88–95
    and Platonic solids 96–101
    qualities 95
colours 170–179, 218
    of the Classical Elements 203, 207, 212, 214
    combined with shape and Elements 220–223
    cool 222, 223
    and the Gateway to Becoming 225
    and music 180–187
    and shape 218
    symbolic qualities 177–179, 224
    warm 220–221, 223
community 20, 22, 44

compass 15, 19, 184
completion 40, 41
concentration 77
    of energy 53
concentric circles 21, 23
consciousness 13, 31, 37, 60, 79, 80, 82, 85, 86, 87, 95, 115, 116, 145, 183, 193, 197, 204, 229, 230
cooperation 50
Copernican system 105
cosmic
    cycles 77
    vibrations 143
    womb 37
cosmological constant 50, 123
cosmos 6, 129, 143, 198
counterpoint 151
creation 34, 39, 40, 41, 45, 46, 61, 74, 90, 129, 143, 145, 152, 160, 185, 186, 200, 205, 216, 229, 230
creativity 17, 50, 59, 81, 96
Cross 28–33, 45
    Rosicrucian 60
crystals 21, 38, 107, 109, 206
cube 10, 28, 29, 84, 85, 96–100, 101, 107, 114, 116, 168, 201, 203, 217
cycle 21, 31, 36, 56, 113, 129, 157, 184, 203
    oestral 207
    of the Universe 129
Cymascope 141, 147
cymatics 139, 140, 141, 190, 226

D

decad 40, 58
decagon 59
decimal system 102, 167
degrees of freedom 123
diamond 35, 106, 107
dimensions 20-21, 112, 118, 120, 121, 123, 127, 128, 163, 196
"dinergy" 180
Directions 28, 30, 33, 37, 43, 45, 54, 93, 95, 111, 219, 224, 225, 229
    cardinal 93
DNA 51, 52, 153, 166
dodecahedron 10, 57, 96, 97, 98, 101, 128
dolphins 134, 137

doors 26, 31, 32, 33, 43, 85, 229
dragon 92, 93, 186, 204, 211
"dragon curve" 71
drum 135, 141, 145, 148, 157
duality 13, 24–27, 40, 45, 59, 74, 223, 225
Dzogchen 94

E

E8 polytope 123–124
E8 model 140
ears 133, 134, 142
earth 201–203 see also Classical Elements
earth womb 201
Edgar Cayce Readings 187
egg 23, 36, 51
ego 31, 45, 60, 81, 82, 85, 211
Egyptians 14, 34, 39, 56, 60, 82, 84, 93, 96, 141, 143, 144, 175, 187, 210
electromagnetic
    force 169
    spectrum 169, 170, 175
    sphere 163, 164
electromagnetism 111, 136, 160, 161, 165, 166
electrons 27, 103, 105, 160, 162, 164, 166, 191
Elements 47, 88, 146, 229
    astrological 99
    chemical 102, 104
    Chinese 88, 93, 94, 218
    Classical see Classical Elements
    combined with shape and colour 220–223
    and the Gateway to Becoming 225
    Native American 95
    Periodic Table of 51, 104, 187
    Tibetan 94
Elixir of Life 91, 93
elliptic geometry 117
emotions 13, 31, 33, 86, 95, 148, 149, 156, 175, 181, 206, 219, 221, 222, 230
emptiness 146, 147
energy 17, 31, 32, 36, 44, 51, 79, 95, 107, 110, 111, 116, 118, 120, 195, 216
    dark 51, 165
    fields 133, 169
    and geometry 140
    grid 166
    light 160

origin of 167
  reverse of time 119
  transmutability 77, 90
  vortices 129
enlightenment 196, 211
ennead 38
enneagram 39
Eternal Return 129
eternity 44, 113, 116
ether 143, 213-215 *see also*
  Classical Elements
Euclidean geometry 117, 118
  evolution of the whole 77, 78
existence 44, 81, 116
Expanding Universe 50, 132,
  164
eyes 81, 84, 86, 87, 115, 170,
  193, 195
  cone cells 172, 173, 174
  rod cells 172, 173
Eye 44, 81, 82, 84, 85, 87, 231
  All-seeing 84, 221
  Dual 81
  Here on Earth 84
Eyes of Horus 82, 84

F

fathom 155
fertile number 37
fertility 39, 41, 48, 53, 54, 56
fertilization 45
fetishes 219
Fibonacci Series 53, 55, 125, 152
fields/field theory 110, 111, 143
fire 208–212, 221 *see also*
  Classical Elements
fire altars 212
flames 196, 210, 211, 212
flow form 206
fluorescent light 162
folding 80
food pyramid 75
force field 24, 163
forces 110, 111, 117, 120, 123
form/forms 6, 36, 55, 106, 110,
  111, 120, 125, 139, 146, 165,
  174, 180, 192, 196, 201, 204,
  208, 216
fourth dimension 118
fractals 66–71,103, 118
  and fractal geometry 66–71
freemasonry 57, 84
frequency 37, 111, 140, 142,
  143, 153, 161, 164, 169, 175,
  181, 187, 190
frieze group 185
fullerenes 106

G

galactic morphology 154

galaxy 21, 47,154, 164
Gateway to Becoming 224–231
  first stage 43
  second stage 199
Gateway to the Heavens 6, 7, 15,
  42, 85, 112, 168, 197–200,
  217, 220, 224, 229, 231
generative
  phi proportions 55
  power of opposites 52
  spirals 53, 71
geomancers 15, 19
Geomatria 41, 143
geometric
  code 23, 132, 147, 160, 183,
  226
  model and symbol 42
  shapes 11, 23
geometry in multiple dimensions
  112
glide reflection 175
glyphs 60
Goddess Energy 203
Goddess Venus 56
golden flower 82
Golden Mean/Ratio/Section/
  Triangle 52, 55, 57, 99, 151
golden rectangles 97
graphite 106
gravitational field 164
graviton 121
gravity 50, 111, 118, 136, 165,
  183
Greeks 38, 56, 58, 88, 89, 93,
  96, 99, 123, 143, 206, 208,
  210, 214
grids 6
Grid of Being 40, 41, 43, 45, 75,
  81, 205, 208, 212
Grid of Life 112, 114, 115, 205
Grid of Space 28–30, 43, 45, 92,
  100, 101, 114
Grid of Time 20, 21, 30, 43, 45
  growth and development,
  sequence of 185, 187

H

halo 145
harmonic
  motion 154
  progression 150, 152
  reciprocals 165
harmonics 137, 139, 149, 151
harmonious coexistence 24
harmony 150, 153, 154, 180
  celestial 150, 155, 156
  and musical notes 152
heart 32, 33, 34, 60, 82, 149
heat 100, 107, 110, 160, 161,
  200, 208, 209

Hebrew
  Ancient 38
  letters 41, 60
helix 51
  double 51, 52, 223
heptagon 184, 185
heptagram 184, 185
heptatonic scale 156
heraldry 39, 92
hexagon 37, 38, 40, 56, 62, 63,
  84, 101, 107, 109, 167, 168,
  205, 208
hexagram 10, 39, 43, 114, 115,
  226
  grid 167
hierarchies 39, 75, 77, 108, 110,
  197
hieroglyphs 141
Higher Self 60, 133, 196
Hindus/Hinduism 79, 87, 88,
  143, 144, 177, 178, 185, 186,
  187, 195, 198, 200, 206, 214
hologram 190, 191, 192
holographic
  brains 192
  fields 196
  light 195
  model 197
  reality 195
  seed 230
Holographic Universe 190, 196,
  224
"holomovement" 192
Holon 20, 21, 23, 31, 32, 50,
  74, 85, 86, 91, 103, 113, 139,
  163, 198, 225, 226
honeycomb 38
HU 147
Hubble Constant 50
Hum, The 135
  Taos Hum, The 135
humming 142
hydrogen 27, 102, 103, 106, 107
hypercube 125, 126
hyperspheres 127

I

ice 109, 205, 206, 210
icosahedron 96, 97, 98, 101,
  128, 206, 214
ideograms 100, 142
Illumination and intention of the
  Mind 193
imagination 10, 11, 86, 119,
  197, 206, 214, 216
Inca 14, 210
incandescent light sources 160
infinite complexity 68–69
infinity 44, 51, 55, 59, 79, 116,
  214

inner knowing 13
insight 13, 14, 82, 86, 195, 214
intent 14, 32, 33, 80, 93, 110,
  135, 194, 196, 216, 217, 219,
  228, 231
intuition 13, 17, 84, 86, 95, 113,
  193, 197, 206
irrational numbers 53
Islam 58, 143, 185

J, K

Jacob's Dream 75
Jain philosophy 85
Japanese 14, 88, 226
journeys 30, 31, 45, 49, 78
Julia Sets 67, 70
Kabbalah 16, 75
knots 155
Koch Curve 70

L

labyrinths 47
Language of numbers and their
  shapes 10–17
language and numbers 143
larynx 135, 142, 147
laser 190
Latin Cross 126
lemniscate 51
Levels of Being 39
life cycles 77, 78
life paths 30
light 19, 160–169, 227
  colours of 170–179
  cone 163, 172
  harmonics 166
  refraction 160, 227
Light Beings 160–169
Lissajous Curve 155
Locating the Centre 85
logic 17, 79, 81, 86, 197
luminescence 162

M

magic and mysticism 57, 93
magic number 185
"magical triangle" 212
magnetic
  field 103, 138, 161, 162, 194
  forces 117
  poles 27
  rope 138
  mana 49
mandalas 7, 16, 198, 200, 229
Mandelbrot Set 62, 67, 192
"Mandelbulb" 70
mandorla 26
mantra 143, 146, 187, 226, 228,
  229
Matrix of Space-Time-Being

42–45, 46, 49, 66, 71, 85, 110, 111, 112, 116, 120, 128, 154, 167, 183, 196, 197, 198
matter 102–111, 118, 119, 129, 166, 210
dark 164, 165
transformation of 90
Maya 102
Mayans 39
Medicine Wheel 16, 90, 95, 219
melody 149, 156, 157, 181
metal *see* Elements, Chinese
Mexico 211
mind, body and spirit 79
Mind scope 193, 197
Mind and sentience 79–87
molecules 51, 102, 103, 107, 133, 161, 163, 167, 183, 190, 205
monad 19
M-Theory 121
music 135, 137, 141, 152, 156
African 157
Chinese 156
and colour 180–187
Indian 157
ratios 153
rhythm and harmony 148–157
and tone colours 156, 157

N

Native Americans 15, 39, 92, 133, 177, 178, 201, 210, 214, 219
Navajo sand paintings 145
n-cubes 125, 126
nefish 49
neutrinos 165
neutrons 27, 103
Nirvana 177, 183, 196
nitrogen 106, 107
n-simplexes 125, 127
n-spheres 125
numbers
language of 10–17
"parents" of 27, 39

O

octahedron 96, 97, 98, 100, 128
Om *see* AUM
orgone 49
Ouroboros 48, 93
ovum 20
oxygen 102, 103, 106, 200, 209, 213, 214
ozone 160

P

parallel lines 117, 120
particles 103, 108, 123, 160,

164, 165, 191
passion flower 59
pattern-forming 209
pattern-sharing 53, 62–71, 74, 118, 175
pentacle 55, 57, 94, 217
pentad 54
pentagon 10, 26, 54, 56, 58, 101, 128, 151, 153
pentagonal star 55
pentagram 54, 55, 56, 57, 58, 59, 60, 151, 153, 217
Pentagrammaton 57, 60
pentaskelion 49
pentatonic scale 156
perfection 39, 91
of Being 45
periodicity 63, 163, 164, 190
perpetual
rhythm 24
transmutation 108
Persians 210
perspective 29, 217, 229
petroglyph 18, 48, 49
phase boundaries 109
phase diagram 109
Phi Ratio 52, 53, 55, 59, 71, 99, 101, 153
Philosopher's Stone 91, 92
phoenix 211
"phonon" 163
photons 121, 160, 165
photosynthesis 169, 213
pineal gland 83
pinnacle 76, 77, 78
Planck Constant 168
Planck Lattice 167
Planes of Being 39, 74, 75, 77, 81, 88, 154, 196, 214, 228, 230
planets 101, 102, 105, 111, 118, 151, 186
Platonic Solids 96–101, 203, 207, 211, 214
pneuma 49
polygons 52, 63, 97, 184
polyhedrals 96
polymorphism 106
polytope 123, 124, 125
prana 49, 214, 222
precognition 195
prefrontal cortex 83, 86, 194
pressure changes 133, 139, 163, 213
Prime Cross 119
Prime Number Cross 167
prime number series 167
proportion 71, 96, 99, 141, 150, 151, 155, 218
protons 27, 103

Ptolemaic system 105
Pulsating (or Inflating) Universe Theory 129
pulse 149, 166
purity/purification 204, 206, 211
pyramid 55, 84, 85, 100, 211, 217, 221
Pythagoras 15, 40, 71, 96, 99, 150, 152, 155
Pythagorean
influence 182
numbers 167
proportion 141
studies 122
theorem 71
Triangles 55
Pythagoreans 58, 90, 91, 185

Q

quantum physics/theory 75, 120, 123, 124, 126, 140, 160, 164, 167, 190
quintessence 88

R

raga 157, 186
rainbow 224, 227, 228
reflection 64, 64, 175
relativity 110, 118, 120, 125, 165, 166
repetition 63
resonance 151, 156
rhombus 35, 37, 38, 115
rhythmic vibration 154
rhythms of life 149
ritual 93, 94, 95, 135, 157, 219
Romans 14, 210
Rose and Cross 60
rotation 64, 67, 175

S

sacred
art 17, 48, 197
fire 210, 212
geometry 14–15, 16, 17, 216, 218, 224
letters/texts 143, 144
sound 142–148
space 198
universal models 16
salamanders 211
salt 201, 206, 212
samsara 45
scale variance 66
scaling 64, 118, 120, 125
seasons 94, 95, 113, 149
seed mantra 229
seed vibrations 229
self 31, 85, 94
Inner Self 32, 46, 93, 179

self-development 45
self-organization 208
self-replicating growth 53, 152
self-similarity 66
self-sustaining systems 209
sentience 44, 81, 86, 87, 94, 113, 114, 116, 120, 128, 129, 231
serpents 48, 52, 129, 226
shapes
combined with colour and Elements 220–223
geometric 23, 228
language of 10–17
and patterns 12, 13, 14, 16, 17, 26, 111, 120, 139, 142, 216
sharing 21, 50, 51, 62, 64
Sierpinski Triangle 70
silence 146, 147, 173, 187, 219
singing 135, 142, 143, 156
singularity 27
Siva Nataraj's dance 145
snakes 48, 134
snowflakes 62, 205
Solar Cross 49
solar year 55, 56
Solfeggio Frequencies 153
sonic bubble 141, 163, 190, 191
sonic spheres 141
sound 16, 87, 111, 130–157, 180, 181, 182, 183, 187, 190, 214, 228, 230
pressure 163, 164
sacred 142–148
silent 133
symbol 144
as a vehicle for geometry 132–141
waves in Nature 133
Space, Aspect of 112, 114
Space-Time 28–33, 81, 87, 101, 110, 113, 120, 121, 123, 165, 168, 191
speed of light 165, 167, 168
sperm 51
spheres 19, 97, 99, 107, 118, 124, 127, 141, 163, 187, 217
Spheres of Heaven 40
spirals 46–53, 66, 118, 129
of the Universe 129
spires 76
Square, Space-Time and the Cross 28–33
Squaring the Circle 31, 58
Sri Yantra 6, 7, 200, 227, 228, 229, 230
star 55, 56, 57, 217
Star of Bethlehem 57
Star of David 37, 85, 100

states of matter 108, 109
stillness 77, 132
String/Superstring Theory 120–121, 123, 190
subtle bodies 186
subtle energies 187
Sufi tradition 147
Sumerian words 27
sunlight 160
swastika 198
symbols 6, 75, 93, 100, 101, 197, 215, 216
   air element 214
   alchemical 99
   Cross 32
   and deeper truths 13
   dragon 92
   earth element 203
   feminine 142
   ideologies 58
   infinity 51, 59, 214
   Matrix of Space-Time-Being 42
   of the original source 144
   pentagon 56
   redemption (Rose and Cross) 60
   Sri Yantra 200
   Sun 20, 32, 33, 74, 81, 196
   T'ai Chi see T'ai Chi symbol
   Tetraktys 40
   Triple Goddess 77
   variety of 14
   water element 207
symmetry 25, 59, 60, 62, 63, 64–65, 67, 123, 141, 185, 218
synaesthesia 182

T
T'ai Chi symbol 13 24, 26, 51, 61, 129, 150, 173, 223, 225
Tantric philosophers 143
Taoists/taoism 26, 39, 93, 94, 95, 96, 144
tattvas 218
telepathy 196
temples 16, 49, 75, 99, 204
tension 24, 27, 165
tessellation 52, 62
tesseract 125, 126
tetrachromacy 174
tetrad 123
Tetragrammaton 41, 60
tetrahedron 96, 97, 98, 100, 101, 107, 114, 127, 211
tetraktys 40, 41, 58, 61, 90, 91, 121, 122
Theory of Everything 120, 123
thermodynamics 125
Third Eye 82, 84, 86, 87, 115, 230
thought, directed 194
thought-form 197, 198, 217, 226
Tibetan philosophy 94
tides 207
Time, Aspect of 112, 113
time, dimensions of 20–21, 113
time zones 22
tornado 53
toroid vortex ring 129
toroidal hexagram 125, 126
translation 64, 65, 67, 175
Tree of Life 75
Triangles and Being 34–41

triangular mountains 76
trinity 35, 36, 39, 40, 41, 46, 78, 79, 84, 115, 121, 209
Triple Goddess 77
triskeles 49
Tychonian geocentric system 105

U
unified field 110, 165
   theory 123
unity 23, 24, 75, 78, 91, 92, 136, 195, 229
Universal
   Energy Field 169
   Plan 31, 45,
Upanishads 186

V
Vedas 185
vesica piscis 24, 25, 26, 29, 35, 36, 37, 61, 81, 82, 92, 115, 142
vibrations 37, 39, 44, 45, 74, 77, 111, 114, 120, 129, 135, 139, 141, 143, 146, 154, 168, 180, 187, 228
visible light spectrum 170
vision 84, 173
Vitruvian Man 58
vocal chords 147
voice 137, 156, 157, 190, 191, 222
vortices 129, 165, 186, 208

W
water 108, 109, 204–207 see also

Classical Elements
waves/wave forms 111, 120, 132, 133, 136, 137, 139, 141
   in light 160, 161, 164
wave equation 136
wave field of light 190
wavelength 161, 164, 170, 173, 181
whale song 137
Wheel of Life 33, 46
wheels 33, 45, 186
   colour 171, 181
whirls/whirling 111, 138
Wiccans 57
womb of creation 26
wood see Elements, Chinese
World Tree 146
Wu Chi 26

Y
Yantra 6, 7, 16, 198, 200, 217, 218, 226, 228
   colours of 227
   see also Sri Yantra
Yin and Yang 14, 26, 93, 129, 223
yoga 79, 198, 217, 219
yoni 26, 142

Z
Zen calligraphy 144
Zen diagrams 218
zero 18, 58
Zodiac 56

Picture credits

Geometric line drawings are by Karen L. French or from the public domain unless otherwise credited below. Part divider images are credited as necessary where they appear in the main text. Photographs, artwork and illustrations are from Jupiterimages ©2011. Other sources are credited below or listed as a public domain image.

National Aeronautics and Space Administration (NASA)
p12: Artist's impression of the extra-solar planet HD 189733 b; Spiral galaxy
p19: Surface of Moon
p20: Eye of the storm
p21: Planetary nebula
p26: Earth-Moon system
p47: Low-pressure spiral cloud formation over Iceland; Interacting spiral galaxies (Hubble); Spiral storm; Clouds off the Chilean coast showing "von Kármán vortex street"
p50: Galaxy NGC 5584

p68: Radar image, TEIDE volcano, Canary Islands, Spain; Lena River delta satellite image – USGS EROS Data Centre Satellite Systems Branch
p69: Omega Nebula
p77: Giant galactic nebula NGC 3603
p102: Crab Nebula was created for NASA by Space Telescope Science Institute and for ESA by the Hubble European Space Agency Information Centre under Contract NAS5-26555
p105: Planets and asteroid belt
p117: Fantasy planet with elliptic magnetic poles
p132: Big Bang noise chart
p133: Satellite image of wave cloud, Amsterdam Island, southern Indian Ocean
p138: Magnetic rope; Heliospheric field of the Sun
p154: Galactic morphology: Hubble "tuning fork" system
p161: Extract from the chart of the Electromagnetic Spectrum
p162: Orion Nebula
p164: Dark matter in a galaxy cluster in the inner region of Abell 1689

(2001), Hubble's Advanced Camera for Surveys
p165: The Cosmic Web (ESA and E. Hallman - University of Colorado, Boulder); Pie charts showing the distribution of dark matter and dark energy
p175: Kepler's Supernova Remnant by Hubble Space Telescope, Spitzer Space Telescope, and Chandra X-ray Observatory
p194: "Eye of God" or helix nebula
p209: Flame Nebula

Robert Webb's Great Stella software
www.software3d.com/Stella.php
p57: Great stellated dodecahedron, p126: Hypercube, p127: Hyperpyramid, p128: Pentagon, octahedron, dodecahedron and icosahedron in 4 dimensions

Other credits
p11: L'Atmosphere from Météorologie Populaire (1888) coloured-in version by Hugo Heikenwaelder, Austria
p70: Evolution of the Greek Cross

fractal by Robert Dickau
p 71: Pythagorean Tree fractal by Guillaume Jacquenot
p98, p100, p101, p203, p207, p211, p214: Platonic solids released by PDD at the wikibooks project
p104: Periodic Table in spiral format by Jan Scholten; Periodic Table traditional layout by Simon Gjerløv from Bitfrost Interacive, Denmark
p106: Carbon allotropes by Michael Ströck
p136: Atmospheric gravity waves
p147: Cymatic image of OM Cymatic image contributed by John Stuart Reid/CymaScope.com
p157: Melakarta ragas hashing method by Basavaraj Talwar, Mysore/Bangalore, India
p172: HSL and HSV diagram by Jacob Rus (2010)
p229: Ceremony for closing the Hevajra Mandala – permission granted for PD use by His Holiness Dagchen Rinpoche, Tharlam Monastery of Tibetan Buddhism, Boudha, Kathmandu, Nepal.

Picture credits and acknowledgments

Public domain

p11: Petroglyphs in Morocco, France, Cholpon-Ata, Colombia, Venezuela

p15: Title page of *Geometrae practicae novae*

p16: Big Horn

p17: Sand mandala; Christian Cross; underwater panther; Japanese pedestal; Incan fabrics

p19: Fly ash (United States Department of Transportation); Bali sand spheres; Volcanic bubbles in water

p20: Artist impression of the Milky Way galaxy; Protoplanetary disc

p22: Johnson *Diagram of the World Time Zones from Washington*, from *Geographicus*

p25: Brain cross-section

p26: Chalice Well gardens

p30: Volcanic stone medallion (14th–15th C), Dagestan

p31: Michael Maier's *Atlanta Fugiens Emblema XXI* (1617)

p33: Ramalingeshvara Temple, Rameswaram, Tamil Nadu; Jellyfish (Discomedusae) – Plate 8 from Ernst Haeckels' *Kunstformen der Natur* (1880); Petroglyph of holon and cross

p35: Mikael Toppelius in the old Kempele church (July 2006)

p37: Leningrad codex cover page E (1010 BC); Koran, China (late 16th C)

p38: Giant's Causeway Ireland; Drosophilidae compound eye (by Dartmouth College, UK); Flocon crystal (USA Government); Zavitan River Hexagons

p40: *Rosa Celeste* by Doré (1892)

p41: Tetragammaton (1109) from *Dialogi Contra Iudaeos* by Petrus Alphonsi; Tetraktys with numbers

p46: Campylobacter bacteria (Agricultural Research Service (ARS), USA Dept of Agriculture)

p48: Maori door lintels from an illustration to a Czech-language travelogue (*Pictures from the Southern Hemisphere*, chapter II (1900); Vor Frelser Kirke, Copenhagen, Denmark; Spiral minaret at Abu Dulaf; Coiled snake sculpture at north altar of the Tenayuca pyramid in Tlalnepantla Mexico; Hindu snake and turtle illustration

p51: Archimedean screw, Egypt

p51: DNA double helix; Double coiled serpents from the Deutsche

Fotothek (1603)

p54: Aztec Sun Cross calendar from *Descripción histórica y cronológica de las dos piedras que con ocasión del nuevo empedrado que se está formando en la plaza principal de México, se hallaron en ella el año de (1790)*; Brittle star (Haeckel Ophiodea) from *Kunstformen der Natur* (1904)

p56: Pencils

p57: Pentacle from the Sixth Book of Moses; Crotona Pentagram ring (c. 525 BC) taken from the book *Imagini Degli Dei Antiche* by V. Catari, (1647); Man in a pentagram from Heinrich Cornelius Agrippa's *Libri tres de occulta philosophia* (15th C)

p58: Flags

p59: Passion flower

p60: Luther Seal (19th C); Drawing of Cross combining pentagon and Cross; Boehme-heart black and white illustration; Rosy Cross illustration

p61: Tokay Gecko

p62: Detail from Mandelbrot set; Snowflake, ice crystals (US government)

p65: Boron hydroxide atomic structure drawing (US government); Protein structure; Haeckel Acanthophracta (1887); Nautilus

p67: Mandelbrot set and magnifications

p68: Fractal broccoli

p69: Mountain eruption (USA Geological Survey)

p70: Julia Set; Koch curve generation; Coloured Koch curve snowflake; 3D fractal, mandelbulb; T-Square fractal; Sierpinski pyramid; Pentagonal fractal

p71: Dragon trees (computer generated); Dragon tree

p75: Angels' hierarchy (12th C) by Hildegard of Bingen

p77: Giant wood wasp life cycle

p78: Phytoplankton (US government); *The Life and age of woman, stages of woman's life from the cradle to the grave* by James Baillie (c. 1848)

p80: Brain coral ball (by Janderk); Brain fungus; Orange puffball sponge (USA National Oceanic and Atmospheric Administration)

p84: The Eye of Cao Dai (by Nyo); Eye of Providence (early 18th C), Beschreibung

p88: Title page of Sir Henry Billingsley's *Euclid's Elements* (1570)

p92: Tripus Aureus (1678) *Theosophie and Alchemie*

p93: *Atalanta Fugiens* (1617) Ouroboros

p97: Single pink crystal; pink rock salt photograph (by Ingo Wölbern); Pyrite photograph (by Anescient); Crystal cell structures

p99: Kepler's Platonic solid model of the Solar system from *Mysterium Cosmographicum* (1596)

p104: Circular table of elements

p105: Tychonian geocentric system (by Fastfission); Mars Earth orbit path (by Tomruen at the English Wikipedia project); Planets paths from *Encyclopaedia Britannica* (1771)

p108: Diamond (United States Geological Survey); Coal from *Minerals in Your World* project, United States Geological Survey and the Mineral Information Institute

p109: Hexagonal ice Ih (by Materialscientist); Triple point of water graph

p115: 3D cube incorporating a star tetrahedron

p117: Hyperbolic triangle; Hyperbolic icosahedron; Non-Euclidean geometry diagrams; Radiolarian from *Kunstformen der Natur* (1904), plate 71, by Ernst Haeckel

p121: Calabi-yau

p124: E8 family graph (larger image) (by Tomruen); E8 model (smaller image) (by Jgmoxness)

p126: Unfolded tesseract (by Dmn)

p126, p127: Higher dimensional graphics

p134: Bat (USA Fish and Wildlife Service)

p137: Standing waves

p138: Magnetic whirls around a wire, *Popular Science Monthly* (1920s); Whirling dervish

p139: Compressed air shells; Drum harmonics 2nd Bessel

p140: Chladni bow method black and white illustration; Chladni figures (1803); Chladni plates

p141: Symmetry of vibrating fluid (free art licence); ESPI vibration in water

p144: All texts; Stone rubbing of Wang Xizhi manuscript (AD 356)

p150: *Harmony before Matrimony* by Js. Gillray (1800–1810)

p151: Kissar by Andreas Praefcke;

16th C counterpoint music

p154: Early harmonic motion equipment

p155: Lissajous knots

p156: Page from *A New Theory of Music Tones* by Zhu Zaiyu (1536–1611)

p160: Flame on earth and in zero gravity

p162: Light sculpture; Panellus Stipticus; Luminescent jellyfish; Zooplankton and Bathocyroe fosteri (deep sea comb jellyfish) (USA National Oceanic and Atmospheric Administration)

p163: Light cone

p170: Light prism

p171: Newton's colour wheel (1600s); Moses Harris' elaborated colour wheel (1600s); 8-colour wheel; Lichtenburg colour Triangle (1775)

p172: Spinning top colour mix; Bezold Farbentafel colour wheels (1874); Philip Runge's colour as three-dimensional wheel (1810)

p180: Newton's colour/tone wheel (1600s)

p184: Stupa/pagoda

p192: Hologram of field mouse

p195: Robert Fludd's diagram of the mind

p198: Sri Meru Yantra produced by the Devipuram temple, Andra Pradesh, India; Mandala gross

p200: Sri Yantra embossed on copper; Sri Yantra 3-dimensional model

p206: Chalice Well

p207: Sun-Moon cycles

p212: Blue flame of alcohol burning; Eternal flame; A Srividya homa; Bratabandha Pooja in Nepal; Olympic Flame Vancouver

p218: Sand mandala being made; Tattvas

p227: Contemporary yantra; Light refraction through water droplet; René Descartes rainbow formation

The publisher would like to thank people, museums and photographic libraries for permission to reproduce their material. Every care has been taken to trace copyright holders. However, if we have omitted anyone we apologize and will, if informed, make corrections to any future editions.

Acknowledgments

A big thank you to my husband and family for their ongoing support and encouragement over the years; to Jo and Peggy for the superb presentation executed by Bookworx; to everyone in Watkins Publishing for believing in my work, for their editorial contribution, sales and marketing and expert advice. A special mention must go to those who generously contributed their images to the public domain. A book like this would not be possible without them. Last, but never least, to my readers, as your enthusiasm and interest in my work spurs me onward.